The Wetland Bird Survey 1992-1993:
Wildfowl and Wader Counts

The results of the Wetland Bird Survey in 1992-93

by

R. J. Waters & P. A. Cranswick

with the assistance of

J. T. Cayford, M. S. Pollitt & D. A. Stroud

Published by
British Trust for Ornithology, The Wildfowl & Wetlands Trust
Royal Society for the Protection of Birds
and Joint Nature Conservation Committee
December 1993

Design by Peter Cranswick, The Wildfowl & Wetlands Trust
Collated by The British Publishing Company Limited

© British Trust for Ornithology, The Wildfowl & Wetlands Trust,
Royal Society for the Protection of Birds
and Joint Nature Conservation Committee 1993

ISBN 0 903793 42 3 (BTO)
ISBN 0 900806 19 2 (WWT)
ISSN 1353-7792

Printed by Hindson Print Limited

WETLAND BIRD SURVEY

Organised and funded by   **British Trust for Ornithology**
The Nunnery, Nunnery Place,
Thetford, Norfolk IP24 2PU

The Wildfowl & Wetlands Trust
Slimbridge, Gloucester
GL2 7BT

Royal Society for the Protection of Birds
The Lodge, Sandy,
Bedfordshire SG19 2DL

Joint Nature Conservation Committee
Monkstone House, City Road,
Peterborough PE1 1JY

WeBS National Organisers: Wildfowl - **Peter Cranswick**, The Wildfowl & Wetlands Trust

Waders - **Ray Waters**, British Trust for Ornithology

This report is provided free to all WeBS counters, none of whom receive financial rewards for their invaluable work. Further feedback from BTO and WWT HQs is provided in the form of the *WeBS Newsletter*. BTO members can read about WeBS work in the regular "Shorelines" section of *BTO News*, while a summary of the each season's counts are published annually in WWT's journal *Wildfowl*.

ACKNOWLEDGEMENTS

This book represents the fourteenth combined report of the Wetland Bird Survey, previously the National Waterfowl Counts and the Birds of Estuaries Enquiry (see *Introduction* for further details). It provides a national overview of the count information, collected during 1992-93 and previous years, which is critical to the conservation of waterfowl populations both within the United Kingdom and internationally. It is thus entirely dependent on the many thousands of dedicated volunteer ornithologists who supply the data and to whom we are extremely grateful. The Local Organisers who coordinate these counts deserve special thanks.

We are also grateful to the following people for providing technical assistance, supplementary information and comments on the draft texts: Dawn Balmer (BTO), Bernie Bell (WWT), Mike Bell (WWT), Andy Brown (EN), John Bowler (WWT), Stephen Browne (BTO), Alison Browning (WWT), Ali Buck (JNCC), Marie Callaghan (WWT), Jacquie Clark (BTO), Nigel Clark (BTO), Deirdre Craddock (JNCC), Sue Davies (JNCC), Tim Davis (WWT), Simon Delany (WWT), Julianne Evans (BTO), Richard Evans (RSPB), Claire Forrest (BTO), Tony Fox (GWGS), Rob Fuller (BTO), Ian Henderson (BTO), Steven Holloway (BTO), Baz Hughes (WWT), John Holmes (JNCC), Jeff Kirby (WWT), Rowena Langston (BTO), Alan Law (JNCC), John Marchant (BTO), Carl Mitchell (WWT), Ken Norris (RSPB), Rosie Ounsted (WWT), Myrfyn Owen (WWT), Richard Pettifor (WWT), Simon Pickering (WWT), Carol Powley (BTO), Dave Price (WAS), Eileen Rees (WWT), Mark Rehfisch (BTO), Ken Smith (RSPB), Mark Tasker (JNCC), Derek Toomer (BTO), Tony Warne (JNCC) and Mike Wilkinson (EN). Many amateur observers also provide us with reports of their studies. These are acknowledged in the text.

The map of WeBS coverage (page 14) was produced using DMAP, written by Dr Alan Morton. The *Weather* section was summarised from reports provided by Tim Hull of the Meteorological Office.

The cover painting of Cormorants is by Mark Hulme. Other illustrations in this report are by Joe Blossom, Steve Carter, Mark Hulme, Paul Johnsgard, Daphila Scott, Thelma Sykes and Suzanne Yarnton.

The Wetland Bird Survey 1992-1993:
Wildfowl and Wader Counts

CONTENTS

Important Notices...4

Introduction..5
Progress and Developments..6
Research and Commercial Contracts..8
Conservation and Management..10
Winter Weather in 1992-93..12
Interpretation of Waterfowl Counts..12
Data Presentation...13
Coverage in 1992-93..13
Total Numbers..16
Monthly Fluctuations...22
Indices...24
Species Accounts..28
    Divers...29
    Grebes..30
    Cormorant...33
    Grey Heron..35
    Wildfowl...36
    Rails..72
    Waders...74
    Kingfisher..94
    Additional Species..94
Principal Sites...97
Species Codes...102

References..103
Appendix 1: National and International Importance...107
Appendix 2: Locations of WeBS Sites...110
Appendix 3: Total Waterfowl Numbers in England, 1992-93..115
Appendix 4: Total Waterfowl Numbers in Scotland, 1992-93...117
Appendix 5: Total Waterfowl Numbers in Wales, 1992-93..119
Appendix 6: Total Waterfowl Numbers in the Isle of Man, 1992-93...............................121
Appendix 7: Total Waterfowl Numbers in the Channel Islands, 1992-93......................122

IMPORTANT NOTICES

WeBS is of critical importance in both national and international conservation work and, to this end, others are dependent upon us producing up-to-date and accurate data. Please, therefore, return counts to your local organiser as soon as possible. Remember that "winter" counts (October to March) should be submitted by the end of **March**. Any counts made during "summer" months (April to September) should be forwarded to the Local Organiser at the end of **September**. For estuarine coastal sites, counters are strongly advised to forward each count to their Local Organiser directly after it has been made so that organisers can collate and summarise the results. The rapid return of information helps to ensure that annual analyses are completed on time, the results summarised promptly in this report, and the most up-to-date data used by conservation bodies. It also allows data to be submitted to the International Waterfowl and Wetlands Research Bureau (IWRB) in time for their annual report. Local organisers please remember that all data should be sent to WWT, Slimbridge. Many thanks for your help.

We would be pleased to advise on the positioning of count boundaries so that they can be matched, for example, with SSSI and other statutory site boundaries, hence increasing the applicability of the count data collected. Please ensure to consult one of the WeBS National Organisers before making any changes to the overall site or sector boundaries of your count area. It is necessary for us to be sure of the precise details of such changes and to amend our computer files and maps accordingly. For ease of administration, we recommend that counters should contact Ray Waters at BTO with queries regarding estuarine/coastal sites and Peter Cranswick at WWT regarding inland ones.

Counting divers, grebes and seaducks in coastal waters is fraught with difficulties and we are aware that trying to do so on set dates is rarely successful. For these species, we would very much welcome more opportunistic records, i.e. counts made when the weather and sea conditions are suitable. Such records will allow important feeding and roosting areas to be identified, and will allow their frequency of use to be examined so send in all records of these please.

Please remember that, under WeBS, we are asking observers to record numbers of wildfowl *and* waders at all sites. Also, because of the increasing demand to identify the importance of particular areas within a large site, we are asking observers to provide data at the finest practical level. For example, if you count a complex of gravel pits, we would be most grateful if you could sub-divide the area into smaller groups of pits or even provide pit-by-pit data. In practical terms, this may allow us to monitor numbers on an individual pit that is, for example, a Site of Special Scientific Interest, which would otherwise be "lost" in a total for the whole pit complex. Such fine-scale data greatly enhance the potential for its positive conservation uses. Many thanks for all your help.

Late Summer Survey 1993

The majority of data from this survey have now been received. However, would those observers who have still to return forms to their Local Organiser, please do so as soon as possible. This will allow us to produce a complete report and provide rapid feedback. Many thanks for your cooperation.

Ringing Schemes

While many species of birds are caught and ringed with the traditional BTO metal ring to identify movements of birds, many waterfowl are also ringed with colour ring combinations or "darvic" rings, tags or collars (coloured plastic with large, engraved letters or numbers that can be read at large distances using binoculars or a telescope). As active fieldworkers, WeBS counters are in an ideal position to record sightings of ringed birds. Please send your sightings of marked wildfowl to Carl Mitchell at WWT and of waders to Jane Marchant at BTO.

Local Organisers

An up-to-date list of Local Organisers is provided with this report. Very many thanks to all those who retired during the last year for their invaluable work, and welcome to all newcomers. However, we still need organisers in Central, Co. Down, Meirionnydd (Gwynedd), North Humberside and Co. Tyrone. If you would be willing to coordinate counts at inland sites in any of these areas (or can volunteer someone who might!) please contact Peter Cranswick at WWT.

It is with much sadness that we have to report the death of George Arnold in May 1993. Our sincere condolences go to his brother Maurice, with whom George had organised counts throughout much of west central England, including the West Midlands, Staffordshire, Worcestershire and Warwickshire, for many years. George will be sorely missed.

INTRODUCTION

October 1993 saw one of the most important developments in the monitoring of waterfowl in the UK for many years with the launch of the Wetland Bird Survey (WeBS). This represents the amalgamation of the two previous major schemes for surveying non-breeding wildfowl and waders, namely the National Waterfowl Counts (NWC) and the Birds of Estuaries Enquiry (BoEE). WeBS retains the objectives of the old schemes:
- to obtain population estimates for wildfowl and waders in the UK during the non-breeding season
- to monitor trends in abundance of these populations
- to identify adverse trends at particular sites
- to provide a sound basis for the protection of sites and populations

The new scheme, however, provides a greatly enhanced capacity for forward planning, improved co-ordination of research, analyses, and handling of data, and its use for conservation purposes. In many ways, WeBS represents a logical development of the count schemes in their various guises, which began in 1947. A brief history of the origin and major developments in waterfowl monitoring in the UK is given in Table i.

Table i. Major events in the history of monitoring of non-breeding waterfowl in the UK

Year	Event
1947	Wildfowl counts instigated by the British section of the International Wildfowl Inquiry Committee in response to increasing threats to wetlands
1954	Wildfowl Counts, later the National Wildfowl Counts (NWC), moved to Slimbridge and organised by The Wildfowl Trust
1967	International Wildfowl Census initiated by International Waterfowl Research Bureau (IWRB)
1969	BoEE launched, organised by British Trust for Ornithology, improving coverage of estuarine sites, extending species coverage to waders and geographical coverage to Northern Ireland
1985	NWC scheme extended to Northern Ireland
1989	Combined BTO/WWT recording form introduced for coastal sites
1991	NWC extended to become National Waterfowl Counts, including all waterfowl, including waders
1993	WeBS launched

All past data, collected under the forerunners of WeBS, are now treated as WeBS data and called such in this and future publications. While the methodologies of the schemes are fundamentally the same, there are differences in a few areas and these are highlighted in the text where the collection of 1992-93 data under the NWC and BoEE schemes are incompatible in the WeBS context.

The WeBS scheme is funded by the British Trust for Ornithology (BTO), The Wildfowl & Wetlands Trust (WWT), Royal Society for the Protection of Birds (RSPB) and the Joint Nature Conservation Committee (JNCC) (the last on behalf of English Nature (EN), Scottish Natural Heritage (SNH) and the Countryside Council for Wales (CCW) and the Department of the Environment for Northern Ireland (DoENI)). All four WeBS partners take an active role in the planning of the scheme and the rolling programme of analyses that use WeBS data. The day-to-day running of the scheme is the responsibility of the two National Organisers based at the BTO Headquarters, Thetford, and WWT Headquarters, Slimbridge.

The success and growth of the count scheme in its various guises accurately reflects the enthusiasm and dedication of the several thousands of volunteer ornithologists throughout the UK who participate. Counts are made at a wide variety of wetlands including lakes, lochs/loughs, ponds, reservoirs, gravel pits, rivers, freshwater marshes, canals, estuaries, sections of open coast and other coastal habitats. Counts are conducted once per month, normally on pre-selected dates. With the aim of monitoring non-breeding waterfowl, September to March are identified as priority months for counting, although pre-selected dates are provided for all months should observers wish to continue their counts throughout the year - these additional data are always welcome.

Additional surveys of other species, principally geese and swans, that are difficult to monitor accurately by the once-monthly counts alone, are also conducted by volunteer counters, organised by WWT. The results of these surveys are used in this report to complement WeBS data where relevant (see also *Progress and Developments*).

A new recording form was introduced in October 1993 for the collection of WeBS data. This allows standardisation of data collected from all sites and, although basically similar to the previous forms of the NWC and BoEE, it asks observers to record information relating to count accuracy and disturbance in a slightly different way with a view to making greater use of these data when analysing the counts.

The launch of WeBS has been taken as an opportunity to make several changes to this report. Firstly, we have changed the name to reflect the new organisation of the count scheme, but more importantly, we have aimed to provide a "seamless" report, with all waterfowl treated sequentially under each of the various headings as an improvement over the previously separate wildfowl and wader sections. We hope this meets with the approval of counters and other readers of this report and welcome any comments and suggestions for further improvements.

PROGRESS AND DEVELOPMENTS

Staff changes

There were no changes in BTO staff directly involved with WeBS during the year. However, the Waterfowl Monitoring and Wetland Ecology Unit of the WWT at Slimbridge, which includes those staff involved with WeBS, saw many changes in its staff complement as part of wider changes within WWT generally in 1992 and 1993. On the departure side, André Gilburn returned to Nottingham University to research the genetics of seaweed flies, and Marie Callaghan left to become a mother. Our thanks and best wishes to both. Richard Pettifor joined in summer 1993 as a Senior Research Officer. Richard comes to WWT with a strong statistical background and experience in life history theory, particularly reproductive decisions in tits (Parus spp.). Richard has taken a key role in WeBS, especially in directing and undertaking the programme of research that arises from the WeBS data. Mark Pollitt joined the Unit in October 1993 as an assistant to the WeBS National Organiser (Wildfowl) and will assist with the day-to-day running of the scheme. Ian Stenhouse also joined the WWT in October and, based in Scotland, will greatly enhance our monitoring and research into geese in that area, especially Pinkfeet and Icelandic Greylags. Jeff Kirby, who, as many of you will no doubt remember, has organised both the BoEE and the NWC in recent years, was appointed Head of Research at the WWT in November. Jeff will consequently be less involved in WeBS issues but, as one of the key players in establishing WeBS, he will be keeping a close eye on developments in the near future. Many thanks to Jeff for his valuable input in the past and (hopefully) the future also.

Low Tide Counts

An important development this year has been the incorporation of the BTO/RSPB National Low Tide Count Programme into WeBS. The main aim of the programme is to collect and regularly update information on the feeding distributions of intertidal waterfowl on the majority of UK estuaries. The main WeBS counts enable the number of each species wintering on estuaries in the UK to be calculated and changes in population levels to be monitored. The WeBS Low Tide Count Programme will augment these data by providing detailed information on estuary usage by wildfowl and waders.

The methods currently being used to collect this information have been developed over a number of years and draw on the experience gained from a variety of short-term studies carried out by the BTO. These standardised methods are applicable to all but the very largest estuaries, where the most distant mudflats cannot be viewed without the considerable danger of observers walking out onto the estuary. Data collection for each estuary is based on pre-established sub-divisions of the intertidal area which can be between 1 and 250 ha in size. A simultaneous complete count is carried out in each winter month (November to February inclusive). Counts are conducted in the period between two hours before and two hours after low tide with feeding and roosting birds recorded separately. Around 10 to 15 estuaries will be counted annually on a rotating basis, thus incorporating all but the smallest and very largest estuaries at least once every five years. The problems associated with counting waterfowl on estuaries vary according to the estuary's size, shape and habitat composition. In all cases, the advice of the experienced local counters is invaluable in adapting the basic counting procedure to help overcome these problems.

By using methods which sub-divide estuaries in this way, the WeBS Low Tide Count Programme provides data which can be used to assess the relative importance of different parts of individual estuaries for different species. Amongst other uses, this type of detailed information is particularly important as a means of assessing the potential impact of localised activities and developments on estuaries. It will also be used in ongoing site designations and management planning carried out by the country agencies (English Nature, Scottish Natural Heritage and Countryside Council for Wales).

The results so far have highlighted the common factors affecting species distributions on a range of estuaries. They have also demonstrated differences in the use made of estuaries by waterfowl at different states of tide. Golden Plover and Lapwing, for example, were often present on estuaries in larger numbers during the Low Tide Counts than during the WeBS counts carried out at high tide. This is largely due to their habit of roosting on the mudflats at low tide but flying inland to feed during the high tide period. In contrast, the numbers of Ringed Plover recorded during the Low Tide Counts were consistently lower than those recorded by the main WeBS counts due to the difficulty of detecting this small, solitary, well-camouflaged species. These differences reflect ecological characteristics of the species concerned. They should not be used to compare the value of the WeBS Low Tide Count Programme with the main WeBS counts because they fulfil very different roles. Taken together, however, the two schemes have provided new insights into the usage of estuarine systems and their hinterland by waders and wildfowl. They are therefore a valuable tool for conservation as well as applied ecological research.

During the 1992-93 winter, counts were carried out on the Camel, Clwyd, Dengie, Eden, Forth, Hamford Water, Lindisfarne, Montrose Basin, NW Solent, Portsmouth Harbour, Strangford Lough, Swale and Wigtown Bay. Thanks to the enthusiasm and hard work of all those involved, a great deal of vital count data were collected. The continuing support of counters means that the following estuaries will be participating in the programme during the 1993/94 winter: Chichester Harbour, Duddon, Inner Thames, Kingsbridge, Langstone Harbour, Poole Harbour, Strangford Lough, Taw/Torridge and Tay.

Surveys and projects

The behaviour and habits of many of the UK's waterfowl require additional surveys to monitor their populations, including surveys relating to their distribution and habitat requirements. Often, these involve different methodologies, such as dawn or dusk counts at roost sites, or searching non-wetland areas by day, usually for geese or swans. In 1992-93, there were specific surveys of Pink-footed and Icelandic Greylag Geese in October and November (Mitchell & Cranswick 1993) and of native Greylag Geese in the Uists in February 1993 (Mitchell *et al.* 1993). Full censuses of Greenland White-fronted Geese, including birds in Ireland, were undertaken in autumn 1992 and spring 1993 by the Greenland White-fronted Goose Study (Fox 1993a). Censuses of Greenland Barnacle Geese on Islay were undertaken in December and March and an aerial survey of wintering birds in Ireland was made in March. There were also regular counts of the Svalbard population on the Solway Firth (Shimmings *et al.* 1993). Dark-bellied Brent Geese was censused in January and February (Cranswick 1993b) and fortnightly counts of Light-bellied Brent Geese at Lindisfarne were made throughout the 1992-93 winter. Age-counts of arctic nesting geese were also made to assess the often dramatically varying breeding productivity of these birds (e.g. Cranswick 1993a). The results of these censuses are referred to in the relevant tables and *Species Accounts* in this report.

"Special Surveys" are conducted by WWT to monitor wildfowl during the breeding and moulting periods and are the responsibility of Simon Delany. Outstanding data for the 1991 survey of introduced geese were received giving revised totals of 63,581 Canada and 19,501 Greylag Geese (Delany 1993a). A "mop-up" survey of areas not originally covered for the 1992 survey of breeding Shelducks in Britain has resulted in almost complete coverage of suitable areas, and has produced a provisional total of 44,700 adult individuals in early summer.

Ongoing surveys by WWT staff and volunteers saw the sixth year of monitoring breeding Shelduck on the River Severn. Initial analyses show the Severn Estuary has held around 2,500 moulting birds in recent summers, concentrated in Bridgwater Bay. However, numbers at Bridgwater have declined in recent years, with slight increases recorded in other parts of the estuary. The fourth season of midweek counts of wintering wildfowl of the Cotswold Water Park in Wiltshire, Gloucestershire and Oxfordshire was undertaken in 1992-93. A recent report (Delany 1993b) ranks pits according to their importance throughout the four year period and also identifies those used during freezing conditions. These data, plus those from a survey of breeding wildfowl in 1993 (Pickering 1993), have been used as evidence to highlight the conservation importance of this area at a public enquiry as two of the most important pits are subject to development proposals.

Summer and autumn 1993 saw a national survey of moulting wildfowl. WeBS sites throughout the UK were visited at least once between mid July and the end of August to count numbers of wildfowl present in a repeat of the 1985 survey (Salmon 1988). Initial feedback suggests the survey was a big success, with the data received to date rivalling the coverage achieved for winter counts. Preliminary results from the survey are expected in mid 1994. Many thanks to all those who participated.

Continuing surveys by the RSPB saw intensive coverage of three important areas for waterfowl in the UK. The 12th season of counts on the Moray Firth, sponsored by British Petroleum, censused the large flocks of sea-duck that prove difficult to monitor accurately using the standard WeBS counts (Evans 1993). WeBS counts on the Somerset Levels and Moors, under the direction of the RSPB Exeter Office, sustained the exhaustive coverage established the previous winter. Counts of the northern half of Cardigan Bay were undertaken for the third successive season, using a combination of boat-, land-based and aerial counts (Green & Elliott 1993). Where compatible, these data have been incorporated into the WeBS databases, whilst the additional counts of sea-ducks have been cited in the relevant *Species Accounts* in this report.

Work continued on Islay, where WWT have been contracted by SNH since 1991 to investigate the ecological requirements of Greenland White-fronted Geese with reference to the social structure of the population, habitat use and feeding site selection. The study also analysed the distribution of the geese on Islay from 1988 onwards and on the movements of marked birds. Fieldwork by Clive MacKay in 1991-92 concentrated on identifying sub-populations of the geese throughout Islay and describing their distribution. Work in 1992-93 by Steve Ridgill included a more intensive study on a smaller area (within a 10 km square near Loch Gorm) to obtain more detailed information on the ecological requirements of the geese, which would be used in developing a conservation management strategy for the birds on Islay.

Wader Productivity Database

Annual fluctuations in breeding success are likely to have a profound effect on the numbers of waders which winter on British estuaries. For the third year ringers have responded to the BTO's request to record the numbers of juveniles in wader catches in order to assess and monitor the effects of productivity on winter numbers.

International Waterfowl and Wetlands Research Bureau (IWRB)

WeBS data from January counts were, as in previous years, supplied to IWRB for inclusion in the International Waterfowl Census. Through this scheme, IWRB monitors waterfowl populations throughout the Western Palaearctic and, as of 1993, south West Asia also. The annual report (Rose & Taylor 1993) also included counts of waders for the first time. In discussing trends for North-west Europe as a whole, totals for many duck species reached record high levels, in many cases mirroring the recent increases observed in the UK. A cold period in January 1993 resulted in low totals for many wader species in some countries, especially in the Netherlands.

In order to make the best use of available resources whilst still providing accurate and rapid feedback on the fortunes of waterfowl, a sample of key wetlands have been selected for annual monitoring in each country. The "reduced sitelist" thus derived will serve as the basis for calculating annual international population indices. IWRB's UK dataset was refreshed in 1993 with up-to-date waterfowl data for all years of the IWC. Provisional results are expected shortly for the selection of the reduced sitelist for the UK.

Waterfowl Monitoring in the Republic of Ireland

While waterfowl counts are long-established in Great Britain, and have been extended to Northern Ireland for a number of years (see Table i), few regular data are available for the Republic of Ireland. Although there have been two periods of intensive monitoring of wintering birds in the early 1970s and mid 1980s (Hutchinson 1979, Sheppard, in prep.), counts in the Republic have otherwise generally been sporadic. While there have been several co-operative single species studies in the Republic as well as the UK, e.g. continuing surveys of Greenland White-fronted Geese (e.g. Fox 1993a), Barnacle Geese, Brent Geese and recently Icelandic Whooper Swans in 1991 (Kirby *et al.* 1992), the absence of regular monitoring represents an obvious gap in our understanding of the waterfowl populations using the geographical unit of Britain and Ireland. To this end, the WeBS partners, the Irish Wildbird Conservancy, the National Parks and Wildlife Service, the Department for the Environment for Northern Ireland and the World Wide Fund for Nature are currently involved in talks to make the exciting possibility of regular monitoring a reality. We are very hopeful that a monitoring scheme in the Republic of Ireland, which will be called *I-WeBS* and which will work closely with WeBS, will be in place before too long.

Data Requests

WeBS data are in constant demand by a variety of bodies for a multitude of reasons. First and foremost amongst these is conservation, with county and local conservation trusts seeking data for sites within their areas, often for management plans, while bodies such as the RSPB and WWT regularly use WeBS data as the basis for many of their conservation policies. The JNCC and conservation agencies also make frequent use of the data to underpin statutory conservation of waterfowl and wetlands in the UK, including site safeguard and wider countryside measures. January data are provided for IWRB's International Waterfowl Census (IWC), which monitors waterfowl populations on an international scale.

A large number of requests are received in relation to research proposals, often from Universities in relation to MSc or PhD studies, but also from other education establishments for various student projects. The BTO and WWT similarly conduct much research on waterfowl using the WeBS dataset. A large number of requests are also made by the commercial sector, usually via consultancies, often to investigate the importance of sites prior to development proposals, with local government offices also requesting WeBS data for similar reasons.

The demand for WeBS data continues to grow and over 200 requests, involving over 1,000 sites, were serviced by BTO and WWT over the last year. To obtain WeBS data, please contact either Peter Cranswick at WWT (wildfowl) or Ray Waters at BTO (waders) for a data request form and further details.

RESEARCH AND COMMERCIAL CONTRACTS

Research

Waterfowl research at the BTO consists of projects administered and funded through WeBS as well as work conducted on a contractual basis by the BTO Habitats Advisory Unit. Recent contracts have focused on the effect of man's changing use of the estuarine environment. Such studies range from assessing impartially the impact of a development or road scheme close to an estuary, to looking at the effects of disturbance on waterfowl using estuarine habitats. During the past year, members of the Habitats Advisory Unit team have been involved in 15 projects relating to the estuarine environment. Some of the projects have involved collecting new data to answer specific questions, whilst others involved analysing the wealth of information already collected by volunteers and held in the BTO's data banks.

For the past four years, the BTO has been monitoring the area around Cardiff Bay as part of a long-running study to assess any possible effects of the proposed Cardiff Bay barrage. This study is unique in that, if the barrage goes ahead, we will have at least five years of monitoring information prior to construction starting. Such a wealth of baseline information extending over several years is exceptionally rare when looking at the effects of large-scale development. This study will not make any difference to Cardiff Bay and its bird populations, but will mean that when it is completed we will have a much better understanding of the impact of such large developments on bird populations. Even before the barrage is built, there have been considerable developments within Cardiff Bay related to the construction of a new road crossing which have led to the loss of some traditional roost sites. In response two mitigation measures were suggested by the BTO. Firstly, the construction of an undisturbed high tide island within the Bay and secondly, a floating high tide roost site. The island is now being used by virtually all the birds roosting in the Bay during high spring tides. Unfortunately, the raft was not as successful as it was damaged before the birds returned from their breeding grounds (Toomer & Clark 1993).

Over the last 20 years, there has been a considerable increase in man's use of estuaries often resulting in increased disturbance to waterfowl. However, we still do not know how far birds are prepared to move between roost sites in order to avoid disturbance. The Wash is an ideal site to look at the distances that individuals move as there are a large number of roosting locations around the estuary. An analysis by the BTO of the Wader Study Group's wader ringing data archive (commissioned by English Nature) revealed that Dunlin and Grey Plover are remarkably site faithful both within and between years (Rehfisch et al. 1993). This information is vital when it comes to deciding the distance needed between disturbance-free refuges in order to accommodate our important waterfowl populations without forcing them to fly longer distances than they do normally.

Another recent example of man's impact on the estuarine environment has been the introduction of new sophisticated methods of cockle fishing. This has led to a dramatic increase in the amount of cockling on some estuaries. The BTO were asked by Scottish Natural Heritage to investigate past BoEE counts of the Solway estuary to identify changes in waterfowl populations and to see if these were associated with the introduction of large-scale commercial cockling on the estuary (Shepherd & Clark 1993). The study did not show any large scale changes in numbers or distributions of waterfowl but there were indications that cockling had a negative effect on bird populations as a whole on those parts of the estuary where it had been concentrated. These declines appeared to be progressively more obvious after several years of cockling. Much of the bird data needed for this project had already been collected before commercial cockling started on the Solway, illustrating the importance of continually monitoring of all Britain's estuaries. Without the long-term dedication of waterfowl counters such retrospective research would not be possible.

Commercial contracts

The Wetlands Advisory Service (WAS), the consultancy wing of WWT, was established four years ago with the aim of applying the expertise of the Trust staff in wetland issues to win contracts in the commercial sector. Many of the contracts involve intensive surveys of particular areas or waterfowl species and thus use the same or a slightly modified methodology as that of WeBS. Some of the larger contracts undertaken in 1992-93 are summarised below.

1993 saw the completion of two years' work to assess the ornithological importance of the Solway Firth and the River Annan catchment on behalf of British Nuclear Fuels Ltd. These baseline data will be used to assess the possible impact of an expansion to the current nuclear power station at Chapelcross. Many local volunteers assisted with regular high and low-tide counts on the estuary, including goose roosts, whilst night-time work, boat-based surveys of sea-duck and continuous monitoring of birds using key areas throughout the tidal cycle were the responsibility of John Quinn, Liz Still, Phil Lambdon and Mike Carrier. The final report (Quinn et al. 1993) concluded that the Solway may be internationally important for up to 14 waterfowl species and nationally important for a further 22, and identified important sectors within the estuary for roosting turnover suggested that numbers of Ringed Plover and Sanderling, which pass through the site in large numbers in spring, were much greater than the peak count, whilst breeding surveys found significant numbers of Lapwing, Oystercatcher and Redshank. Many thanks to all who assisted with this project over the last two years.

The investigation into the importance of reservoirs for waterfowl in the northeast England region, on behalf of Northumbrian Water plc, entered its second season in 1993. Anne Westerberg, Andrew Donnison and Lys Muirhead, assisted by many of the local volunteer counters, have undertaken comprehensive surveys throughout the region, including a "blitz" of 200 standing waters, 12 major rivers and all coastal areas in January 1993. With the assistance of local anglers, work in summer 1993 has sought to determine the effect of different numbers of fishermen and Northumbrian Water personnel on waterfowl usage of sites.

The waterbodies in south-west London and adjacent areas of Berkshire and Surrey represent one of the most important inland areas for waterfowl in southeast England. However, the strong pressure for development, owing to the high population density and huge infrastructure in this area, presents many potential problems when trying to resolve the issues affecting waterfowl conservation, especially when the habitats used by waterfowl are "working" waterbodies, such as reservoirs, gravel pits or sewage farms. As a result, WAS was commissioned to conduct several projects in this area, which have been largely the responsibility of Mark Underhill, assisted by other Trust staff and many local volunteers. On behalf of Thames Water Utilities plc and English Nature, weekend and midweek counts were undertaken on all waterbodies in the study area during 1992-93 (Underhill & Robinthwaite 1993a). More birds were present during weekend counts compared with midweek ones, notably Tufted Duck, Gadwall and Great Crested Grebe. Further analyses examined bird assemblages with respect to environmental features, and found the numbers and diversity of species to be correlated with the size of the site and the length and complexity of the perimeter, with food availability thought to be the major factor influencing abundance (Underhill et al. 1993). Subsequently, a late summer survey was conducted on behalf of English Nature to examine the populations of waterfowl present at this time. Preliminary results revealed large numbers of several species, notably Shoveler, and showed there to be a large difference in the sex ratios of several species in the study area. Using these data, suites of individual waterbodies were identified which supported certain proportions (from 40% to 100%) of the total numbers of birds for 13 key species. These WAS recommendations will be presented to English Nature for use in statutory site designation.

Further work in the southwest London area, including weekend and midweek counts during winter and breeding surveys, was undertaken for Ecoscope Applied Ecologists, under contract to the Department of Transport, to investigate the possible effects of widening the M25 between Junctions 12 and 25. The study area was found to support 30% of the waterfowl within the southwest London proposed Special Protection Area, with Wraysbury Gravel Pits, Thorpe Water Park and Staines Moor being particularly important areas (Hill *et al.* 1993). Nocturnal monitoring of Staines Moor, using night-vision equipment, found 15 species of waterfowl feeding at the site, including 22 European White-fronted Geese. Further analysis of data for Wraysbury Gravel Pits was undertaken for WS Atkins Environment, funded by Surrey Water Ltd, in relation to the proposed use of two pits as a source of pure water for blending with the River Thames in the event of excessive nitrate levels from agricultural run-off occurring during dry weather (Underhill & Robinthwaite 1993b). Many thanks to the counters who assisted in these surveys.

WAS was also commissioned by the National Rivers Authority (NRA) to determine the numbers and distribution of Goosanders and Cormorants wintering on the lower River Wye and the numbers and distribution of breeding Goosander on the upper Wye. A total of 164 wintering Cormorants were found widely located along the river, being absent from around 50% of the stretches surveyed, and almost 250 Goosanders, females outnumbering males by 1.8:1 (Underhill 1993). Most males had departed for moulting areas in Norway by the time of the summer counts. Nineteen females and 128 juveniles were identified in July. These numbers represent a 153% increase on those recorded by a similar RSPB survey in 1985.

CONSERVATION AND MANAGEMENT

Site designations

Any site recognised as being of international ornithological importance qualifies for classification as a Special Protection Area (SPA) under the EC Directive on the Conservation of Wild Birds (EC/79/409), whilst a site recognised as an internationally important wetland qualifies for designation as a Ramsar site under the Convention on Wetlands of International Importance especially as Waterfowl Habitat. Criteria for recognising internationally important concentrations of waterfowl have been drawn up by the Ramsar Convention Bureau and require a site regularly to support either 1% of the international population of a particular species or subspecies, or a total of more than 20,000 waterfowl of all species (see Appendix 1 for further details). A list of potential SPAs and Ramsar sites in the UK, including those identified for their importance for waterfowl, is maintained by JNCC (see Stroud *et al.* 1990). Until recently, slow progress on the designation of Ramsar sites and SPAs by government had drawn criticism from many bodies within the UK and also the EC Commission. However, more headway has been made in 1993, with the designation of a further 11 Ramsar Sites and classification of a further 15 SPAs since *Wildfowl and Wader Counts 1991-92*. These new designations reflect real progress in the SPA/Ramsar designation programme, and the Department of the Environment (DoE) and Scottish Office are to be congratulated on this achievement. This commitment to the international community to protect these areas serves to strengthen further the protection given to sites under national legislation. A complete list of all sites currently designated under the Directive and the Convention is given in Appendix 1.

Conservation directives and international legislation

The Directive on the Conservation of Natural Habitats and Wild Fauna and Flora (92/43/EEC), or "The Habitats and Species Directive" for short, is important for conservation in the UK and Europe as a whole. The Directive has already been signed by government and a consultation paper has recently been circulated to conservation bodies and other interested parties for comment. However, many conservation bodies feel that the government has missed an opportunity to enhance nature conservation in the UK by not introducing new legislation to accompany the Directive.

Under the Directive, Special Areas of Conservation (SACs) will be designated to afford protection to internationally important sites and, with current and future SPAs, will comprise the EC Natura 2000 network of protected sites. The Habitats and Species Directive provides the potential for significant progress to be made in the conservation of important marine areas, including those areas important for birds. Although the protection of marine areas is required by the EC Birds Directive, it has not until now been possible to fulfil these obligations since the government has linked UK implementation of SPAs to the SSSI mechanism, which only extends to the Low Water Mark. However, if the potential of the Habitats and Species Directive is realised, then we could hope to see an effective mechanism of implementing both marine SACs and SPAs, thereby protecting important offshore areas. Marine habitats have featured prominently in news headlines in 1993, notably because of the *Braer* incident in Shetland. Also, under the 14th round of offshore oil and gas licensing by the Department of Trade and Industry (DTI) in July 1992, 21 licences were granted in environmentally sensitive areas, including areas important for sea-ducks, against the advice of JNCC, their statutory conservation advisors. However, the country agencies are working with government and oil exploration companies to minimise potential impacts in these areas. While no licences have been granted for drilling to take place in these areas, we await the possible outcome of the discovery of large reserves in these areas with justified concern given the apparent DTI approach to environmental issues.

The conservation "buzz-word" of the moment is undoubtedly biodiversity. The UK government signed the Convention on Biological Diversity at Rio in 1992, although it has yet to ratify the Convention. It calls, amongst other things, for governments to conserve biological diversity and ensure the sustainable use of habitats and species, a call which has been recognised in the Habitats and Species Directive obliging the government to "contribute towards ensuring biodiversity through the conservation of natural habitats and wild fauna and flora". A UK Action Plan for biodiversity has recently been published by the government (Anon 1994), while the non-governmental conservation organisations have also published their aspirations for the implementation of the Biodiversity Convention in the UK (Anon 1993). Biodiversity can be expected to feature even more prominently on conservation agendas in the coming year.

Population estimates

Population estimates for wintering waterfowl in Great Britain have recently been revised using WeBS data (Kirby in prep., Cayford & Waters in prep.). With the dynamic nature of many populations, these updates are not only of scientific interest, but are particularly important for conservation, providing the basis by which sites holding nationally important numbers of a species may be identified. New 1% criteria have been derived from these estimates, and are provided in the *Species Accounts* and in Appendix 1. It is planned to revise population estimates on a regular basis in the future to ensure that up-to-date criteria are used in the conservation of Britain's waterfowl. Similarly, the 1% levels for all-Ireland populations of wintering wildfowl and waders have recently been assessed (Way *et al.* 1993). These are given in Appendix 1 and in the Species Accounts as appropriate.

Some international population estimates for waterfowl were also revised in 1993 by IWRB and 1% levels were agreed at a recent meeting in Denmark. These figures have been used throughout this report. International population estimates will be reviewed on a regular basis, with a major revision scheduled for presentation to the 1996 Ramsar meeting.

Introduced species

The UK, as with many other countries, has gained part of its fauna from introductions, both deliberate and accidental, of non-native birds. Many populations have grown to become self-sustaining in the UK and the British Ornithologists' Union has recently assessed the status of Britain's birds, identifying several wildfowl populations to be partly or wholly feral in origin (Vinicombe *et al.* 1993). They point out that the consequences of these introductions have rarely been beneficial, with the threat posed by the Ruddy Duck to the already endangered White-headed Duck a topical example (Hughes & Grussu in press). Although, under both the Wildlife and Countryside Act 1981 and The Wildlife (Northern Ireland) Order 1985, it is an offence to release or allow to escape any animal not regularly occurring in Britain in the wild state, many non-native species, especially wildfowl, continue to occur in the wild (e.g. Delany 1993a). Many of these species already feature in WeBS and data are published in this report and, in accordance with recent papers and county bird reports, we would encourage all observers to record "alien" species in future.

The IWRB, JNCC and DoE jointly organised the International Ruddy Duck Workshop at the WWT's centre at Arundel, West Sussex, in March 1993. Over 50 delegates from 10 countries in Europe and North Africa attended, along with representatives from BirdLife International, the European Commission and the Bonn Convention Secretariat. The meeting agreed a common goal to stop and reverse the population and range expansion of the introduced Ruddy Duck in the Western Palaearctic, in order to safeguard populations of the globally threatened White-headed Duck. The workshop recommended a range of actions to tackle the problems posed by captive and feral Ruddy Ducks.

Cormorants

Cormorants have received much press during the past year, often accused of the depletion of fish stocks in rivers and still-water fisheries. A recent meeting of the Cormorant Research Group in Gdansk, Poland, saw the formulation of a Position Statement (Kirby 1993) outlining, amongst other things, the intention of the Cormorant Group to conduct further research into this conflict. Although a recent WWT survey found densities of Cormorant on the River Wye not to be particularly high (Underhill 1993), and despite there being no proven evidence of Goosanders having a detrimental effect on fish stocks, licences were recently issued by the Ministry of

Agriculture, Fisheries and Food (MAFF), permitting the control of Cormorants and Goosanders on the English section of the River Wye. Although this is the first time licences have been issued for a river in England, licensed killing of sawbills and Cormorants has been occurring on Scottish rivers for many years. Given the lack of scientific evidence against these birds, particularly regarding riparian fisheries, the question has to be asked as to what criteria MAFF and the Scottish Office, Agriculture and Fisheries Department, have applied in deciding that they are causing "serious damage" to fisheries on rivers.

WINTER WEATHER IN 1992-93

The winter of 1992-93 was generally mild with no prolonged periods of sub-zero temperatures. Autumn began cool and unsettled, followed by a series of mild spells with heavy rainfall and occasional cold snaps prior to Christmas. The New Year saw the onset of a stormy period, which in turn gave way to exceptionally dry and mild weather for the remainder of the winter months.

September began as a continuation of the summer's unsettled weather. Temperatures until the 26th were notably cool, with westerly winds bringing showers and periods of heavy rain to all parts of the country. A freak storm on 17-18th rained down 7 cm hailstones on the Foulness area of the Thames Estuary. Over 3,200 dead birds were recovered including over 900 waders and 1,000 gulls. A change to southerly winds for the latter part of the month allowed temperatures to rise once again to normal and above.

In one of the coldest **Octobers** since the war, prolonged periods of unsettled weather dominated. All parts of the country suffered temperatures well below the seasonal norm, with strong northerly winds dominating in the first half of the month and temperatures falling as low as -7°C in Norfolk and the Grampians. The east of the country received above average levels of rainfall, which in turn produced the earliest heavy snow falls in the Cairngorms for 10 years.

In contrast, southerly and south-westerly winds and high rainfall combined to make **November** one of the mildest this century. Continued spells of persistent rain affected many areas, with 28 rivers being placed on full flood alert by the end of the month. Strong winds and gales affected many parts of the country during the middle of the month, particularly northern Britain. Average monthly temperatures were higher than those for October for the first time in over 50 years.

December began unsettled, with strong southerly and south-westerly winds and widespread heavy rain or showers. Northern and western parts of the country were wetter than average, with remaining areas receiving less than average rainfall. From the 17th, a high pressure area provided settled conditions over most places for the remainder of the month. Temperatures stayed below zero for many days in some places, with thick fog patches persisting throughout the day around York and the north Midlands on the 21st.

1993 produced the fifth warmest **January** this century, with almost all parts of the country recording average temperatures 2-3°C above average. The mild temperatures were accompanied by strong winds and gales, producing one of the stormiest months ever in Scotland and forcing the *Braer* oil-tanker aground in the Shetlands. Much of the UK had higher than average rainfall, up to three times the norm in some places, with flooding occurring on the 10th in western and southern England and parts of Wales.

February saw the storms subside and progress into settled, benign weather, even in the north. Everywhere was mild, particularly in northern England and Scotland, temperatures averaging 2-3°C above normal. A wintry spell in the last few days saw many roads blocked by drifting snow in northeast Scotland.

The cold spell at the end of February was short-lived and **March** saw a period of mild, dry and settled weather become established over the country. Temperatures were again above average throughout the country, and only heavy rain on the last day of the month prevented March 1993 being one of the driest months on record.

INTERPRETATION OF WATERFOWL COUNTS

Caution is necessary regarding the interpretation of waterfowl counts and their application, and the limitations of these data, especially in the summary form which, of necessity, is used in this report. The primary aim here remains the rapid feedback of key results to the many participants in the WeBS scheme. More detailed information on how to make use of the data for research or site assessment purposes can be obtained from the appropriate headquarters.

Explanation of the basis for the qualifying levels used for defining both the international and national importance of sites is provided in Appendix 1. In the *Species Accounts and Principal Sites* sections, it is necessary to bear in mind the distinction between sites that **regularly** hold wintering populations of national/international importance and those which may happen to exceed the appropriate qualifying levels only in occasional winters. This follows the recommendations of the Ramsar Convention, which states that key sites identified on the basis of numbers of birds should support such numbers on a regular basis (calculated as the mean winter maxima from the last five seasons in this report). Nevertheless, sites which irregularly support nationally/internationally important numbers may be extremely important at certain times, e.g. when the UK population is high, during the main migratory periods or during cold weather, when they may act as refuges for birds away from traditionally used sites. For this reason also, the ranking of sites according to the total numbers of birds they support (both in the *Species*

Accounts and Table 55) should not be taken as a rank order of the conservation importance of these sites, since certain sites, perhaps low down in terms of their total numbers, may nevertheless be of critical importance to certain species or populations at particular times.

Peak counts based on a number of monthly visits to a particular site in a given season will reflect more accurately the relative importance of the site for the species than do single visits. It is important to bear this in mind since, despite considerable improvements in coverage, data for a few sites presented in this report derive from single counts during 1992-93. Similarly, in assessing the importance of a site, peak counts from several winters should ideally be used, as the peak count made in any one year may be unreliable due to gaps in coverage and disturbance- or weather-induced effects. The short-term movement of birds between closely adjacent sites may lead to altered assessments of a site's apparent importance for a particular species. More frequent counts than the normal once-monthly WeBS visits are necessary to assess more accurately the rapid turnover of waterfowl populations that occurs during times of migration or cold weather movements.

Information collated by WeBS and other surveys can be held or used in a variety of ways. Data may also be summarised and analysed differently depending on the requirements of the user. Consequently, calculations used to interpret data and their presentation may vary between this and other publications, and indeed between organisations or individual users. The terminology used may not always highlight these differences. This particularly applies to summary data. Such variations do not detract from the value of each different method, but offer greater choice to users according to specific requirements. This should always be borne in mind when using data presented here.

DATA PRESENTATION

The format of data presentation follows closely that of the last report. The period covered comprehensively by this report comprises the entire winter (September to March for wildfowl, November to March for waders). Counts of wildfowl made outwith the September to March period have been used in cases where they represent the maxima for the count season (June to May), whilst additional information for waders relating to the spring (April to June) and autumn (July to October) is provided for species with notable passage populations. In *Species Accounts* for waders, non-estuarine coastal sites are identified by an asterisk (*). Wader data for inland sites is provided for the second season (see *Coverage*) and these are identified by a hash (#).

Data derived from sources other than the routine monthly counts are clearly identified throughout, either by means of specific references or by use of a cross (+) to identify counts derived from WWT's goose censuses. The flagging of goose counts in this way is important as such surveys rely on different methodology (e.g. dawn/evening flight counts, field searching) from that adopted in the mid-monthly visits to wetlands. Furthermore, the dates of goose surveys have frequently differed from those used for the standard monthly counts.

In Tables 1 & 2, total counts for all species have been presented except for hybrid and domestic wildfowl. This enables an assessment of the true scale of WeBS monitoring with regard to particular species. In order to save space, the following abbreviations for wetland types have been used for site names in all tables:

Br.	Broad(s)	GP(s)	Gravel pit(s)	R.	River
Est.	Estuary	Hbr	Harbour	Rsr	Reservoir(s)
Fth	Firth(s)	Lo.	Loch(s) or Lough(s)	WP	Water Park

The location of all sites, including all estuaries, mentioned in this report are given in Appendix 2.

COVERAGE IN 1992-93

The priority dates scheduled for monthly winter counts at inland sites in 1992-93 were 13 September, 18 October, 15 November, 13 December, 17 January, 14 February and 14 March, whilst those at coastal sites, timed to coincide with the best tidal conditions for censusing estuarine birds, were 19 July, 16 August, 13 September, 11 October, 15 November, 13 December, 10 January, 7 February, 7 March, 11 April, 9 May and 20 June. From 1993-94, the recommended priority dates will be the same for all sites. As in earlier years, special effort in January was directed at covering as many extra sites as possible to coincide with the International Waterfowl Census organised by IWRB throughout Europe.

Count data from a total of 2,375 WeBS count units (either the site or its subdivisions, termed sectors, used by individual observers when a team is required to count a large site) in the UK were stored individually on computer files 1992-93. This is considerably more than last season, but still fewer than the record total of 1990-91. However, the numbers of units counted in each month were generally higher than in either of the last two seasons in Great Britain, but slightly lower in Northern Ireland. Although more intensive coverage was achieved in some areas, much of the increase is attributable to the computerisation of data from large or complex sites (e.g. estuaries and gravel pits) at the level of their constituent count units.

A total of 1,530 count units were covered in England. Notable county returns included Lancashire (90 sites), Derbyshire (71), Hampshire (70) and Kent (60), the top three counties remaining in the same positions as in 1991-92, while excellent coverage in north Kent elevated the county's position in these rankings. Cumbria, Lincolnshire, Surrey, Humberside and North Yorkshire all returned 50 or more units. The 485 Scottish units counted included large returns

Figure 1. COVERAGE BY 10 KM GRID SQUARES FOR THE WETLAND BIRD SURVEY IN 1992-93.
Small dots represent one WeBS count unit per 10 km square, medium dots represent two units and large dots represent three or more units.

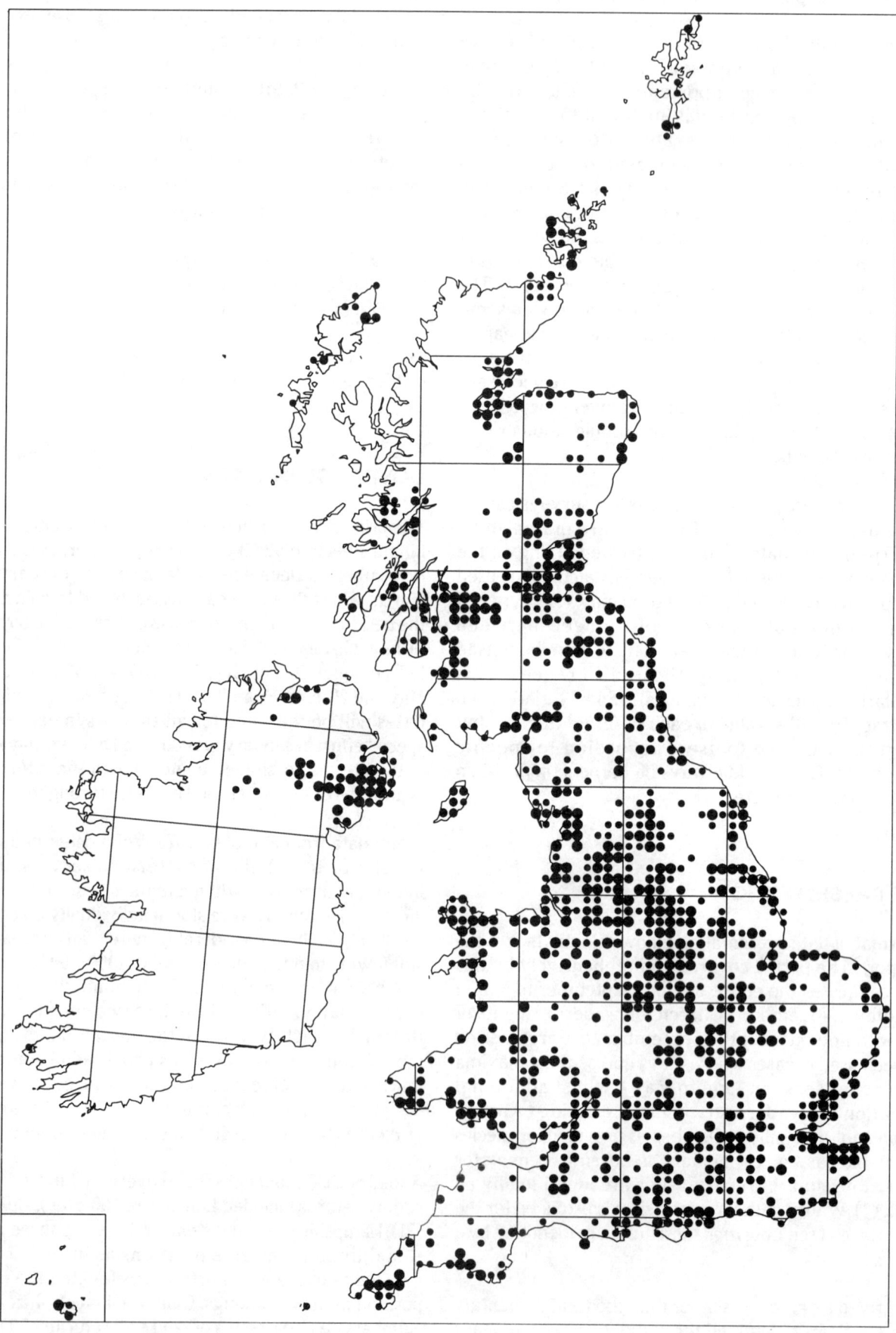

from Strathclyde (118), Highland (69) and Borders (62). The 194 units covered in Wales represented a welcome increase on 1991-92, with returns from Gwynedd (63) and Dyfed (40) the most significant. Only 127 units were counted in Northern Ireland, with County Down (89) contributing the majority of returns. The Isle of Man (11) and the Channel Islands (10) both returned one less unit than in 1991-92.

All of the 129 UK estuaries identified by WeBS (including the 12 sites reclassified as estuaries in last year's report) were counted in 1992-93 with the exception of the Dart, the Looe and the Tyne. Although the total number of waterfowl on all three estuaries combined is less than 1,000, it is important to achieve as complete coverage as possible on all estuaries. Allowing for these three sites, a total of 126 estuarine and over 40 open coast sites were counted in 1992-93. At all of these sites, complete counts were made at least once in the 1992-93 winter except at Carmarthen Bay and Swansea Bay.

In 1992-93, for only the second winter, waders were recorded at around 1,400 inland WeBS sites. This represents a large proportion of all inland sites. However, under WeBS, we are asking observers to record waders at all inland sites to improve our monitoring and understanding of the UK's wader populations.

Several supplementary surveys of geese plus counts of sea-ducks at various sites using non-WeBS methodology were also conducted at WeBS and other sites in 1992-93 (see *Progress and Developments*).

All 10 km squares containing WeBS sites visited in 1992-93 are signified by a dot in Figure 1. The location of a count unit is shown using only its central grid reference. Thus, for example, 11 dots are used in Figure 1 to display the Humber Estuary WeBS site since counts from all sectors are stored individually on computer. Loughs Neagh and Beg are represented by just one dot, even though the site stretches over several 10 km squares, as only the total figures for the whole site are stored on computer.

A total of 1,067 10 km squares contained WeBS count units that were visited in 1992-93, 18 more than in the previous season. Of these, 513 (48%) contained one count unit, 265 (24%) contained two count units and the remaining 300 (28%) held three or more count units. Seven 10 km squares contained 10 or more units visited during 1992-93, with those around southwest London featuring highly as a result of the intense coverage and computerisation of data at the level of individual waterbodies.

The map highlights some gaps in coverage, notably where there is an absence of either human population or wetland habitat. Notable, however, is the lack of data for inland sites in Dumfries & Galloway, which unfortunately were lost in transit. Efforts to obtain another copy were hampered by the Local Organiser moving to Norway. We hope to be able to update the data by the next report.

TOTAL NUMBERS

Tables 1 and 2 show the total numbers of waterfowl recorded by the WeBS scheme in winter 1992-93 in Britain (including the Isle of Man but excluding the Channel Isles) and Northern Ireland respectively. Figures in these tables are derived from the WeBS monthly counts and goose censuses only. Higher totals for certain species (e.g. some sea-ducks) can be calculated by including counts from special surveys made by other organisations and these are highlighted in the *Species Accounts*. The totals for England, Scotland, Wales, the Isle of Man and the Channel Islands are given separately for each area in Appendices 3 to 7.

Numbers of waders are provided separately for estuarine/coastal and inland sites within the tables. This allows comparison of coastal figures with previous reports and also provides some indication of the proportion of each species that utilises inland wetlands. Although waders were counted at many inland sites in 1992-93, coverage is not comparable with that for wildfowl and for this reason also, 1992-93 data for waders at coastal and inland sites are presented separately.

Wildfowl and associated species

Numbers of divers in Britain were similar to those recorded in 1991-92, although they increased in Northern Ireland. Numbers of Great Crested and several of the rarer species of grebe were present in large numbers throughout the UK, although Little Grebe numbers remained rather low. Cormorant numbers were noticeably higher in Northern Ireland and slightly less so in Britain. Generally, fewer Grey Heron were recorded in 1992-93 than in the previous winter. Mute Swan numbers were about average, though migrant swan numbers were well down on 1991-92 levels. Peak counts of migratory geese, as with the swans, were generally low following poor breeding success in 1992, while Canada Goose numbers were also reduced after the high counts in 1991-92. Species of feral or introduced wildfowl, such as Snow Geese and Red-crested Pochard, were recorded in larger numbers, no doubt partly as a result of increased observer interest.

Shelduck, along with several species of dabbling ducks, were recorded in smaller numbers in 1992-93, with particularly large declines of Pintail and Shoveler. Wigeon numbers remained very high, although lower than last season's record totals. Mallard numbers were almost identical to 1991-92 and Gadwall numbers rose slightly. Pochard and Tufted Duck numbers were considerably lower than normal in Northern Ireland, although around average for Britain. Sea-duck numbers were much reduced for several species, although they are notoriously difficult to monitor even under favourable viewing conditions. Red-breasted Merganser numbers were higher throughout the UK, although Goosander numbers recorded in Britain declined. Numbers of Coot recorded in 1992-93 were higher than last season, although around average for recent years, whilst fewer records of Moorhen and Water Rails were received.

Waders

The totals presented in Tables 1 and 2 combined should approximate the total UK population size for those species which are heavily concentrated on estuaries (i.e. Grey Plover, Knot, Dunlin, Black-tailed Godwit and Bar-tailed Godwit) since these are comprehensively covered by WeBS. However, recorded UK count totals for Purple Sandpiper, Turnstone, Sanderling and Ringed Plover will be well below the national population level since WeBS covers only part of the non-estuarine or open coast shores. Open coast waders have been comprehensively counted only in 1984-85 for the Winter Shorebird Count, so a repeat survey is long overdue. Recorded WeBS national totals for Lapwing, Golden Plover, Snipe, Jack Snipe, Ruff and Green Sandpiper have increased greatly as a result of counts at inland sites. However, the absence of data for some sites, means that these totals still represent very conservative estimates on the total national population of these birds. Numbers of the highly cryptic and skulking Jack Snipe and Snipe counted at each site are likely to be far smaller than the populations of those species actually present but for other species the figures should provide reliable estimates of the numbers present on the sites counted. Other wader populations not included in WeBS are the sizeable flocks of Lapwing and Golden Plover that are present on agricultural land well away from any wetland.

Estuaries and coastal wetlands

Tables 1 and 2 show the total numbers of each wader species counted in each winter month of 1992-93 in both Great Britain and Northern Ireland. The numbers of sites (including both estuaries and non-estuarine sites) covered in each month are also given. In the 1992-93 winter, British totals for five wader species reached the highest levels ever reached since the BoEE began in 1969. Three of these record winter counts were made in December 1992, thus pushing the UK all wader total to a new high of over 1,835,000 birds. The main contributory factor was the British count of over 370,000 Lapwing but also responsible were the record counts of Golden Plover and Knot. In Northern Ireland the December counts were typical of recent winters except that particularly low counts of Golden Plover and Ringed Plover were made. The low Northern Irish Golden Plover total was more than quadrupled in the February count, suggesting that birds may have moved in from Britain, where numbers dropped by around 30,000 birds between December and February. The British Lapwing total fell by over 50% between December and January whereas numbers recorded in Northern Ireland rose only slightly. The other species producing record-breaking counts in other winter months were Avocet and Black-tailed Godwit. Both species are concentrated in S Britain and have been wintering in the UK in increasing numbers over the past decade or so. The UK total of Avocet exceeded 2,100 for the first time and over 10,000 Black-tailed Godwit were recorded in a winter month for the first time. This was also true for Northern

Ireland where the January count of 440 Black-tailed Godwit was also a winter record.

In contrast, numbers of Sanderling, Snipe and, in particular, Turnstone recorded in the UK were below the averages of recent winters. Sanderling and Turnstone winter predominantly on open coasts in the UK and many are therefore missed by WeBS counts. Of more concern are the low numbers of Dunlin recorded in both Britain and Northern Ireland.

Inland wetlands

In 1992-93, for the second winter, waders were recorded at inland wetlands. Totals recorded in Great Britain and Northern Ireland are given in Tables 1 & 2, together with the number of sites covered. As at tidal wetlands Lapwing and Golden Plover numbers were high, especially in December. Also in accordance with counts at coastal sites, Dunlin numbers were lower than in 1991-92. In contrast to the coast, however, inland totals of Curlew were well up with the exception of March. For Oystercatchers and Redshank the rise in numbers recorded due to birds moving into their inland breeding sites in late winter was again noted, but no clear peak was recorded for Curlew in March 1993. In Northern Ireland numbers of Lapwing and Golden Plover were surprisingly lower than in the previous winter, although Curlew totals were more than double those of 1991-92. It is relevant, however, that in 1992-93 waders were monitored at around 30% fewer inland sites in Northern Ireland than in the previous winter.

Key to Table 1

+ *Counts include data from the following goose censuses: national census of Pink-footed and Greylag Geese in October and November; December and March censuses of Barnacle Geese on Islay; December census of Barnacle Geese on the Solway; December census of Dark-bellied Brent Geese; November census of Light-bellied Brent Geese on Lindisfarne; international censuses of Greenland White-fronted Geese in November/December and March/April. See Progress and Developments and Species Accounts for more details.*

* *In all months except September, the feral component of this species is approximated by totalling counts from English (excluding Northumberland) and Welsh sites only and adding 2,340 (after Delany 1992) for the feral birds in Scotland. All other birds in Great Britain (apart from the native population in the Outer Hebrides, Coll, Tiree, Colonsay and parts of Sutherland) are considered to be from the Icelandic population.*

** *Total wildfowl represents numbers of all divers, grebes, Cormorant, swans, geese, ducks and rails.*

*** *Total waterfowl represents numbers of all wildfowl (as above), waders and Grey Heron.*

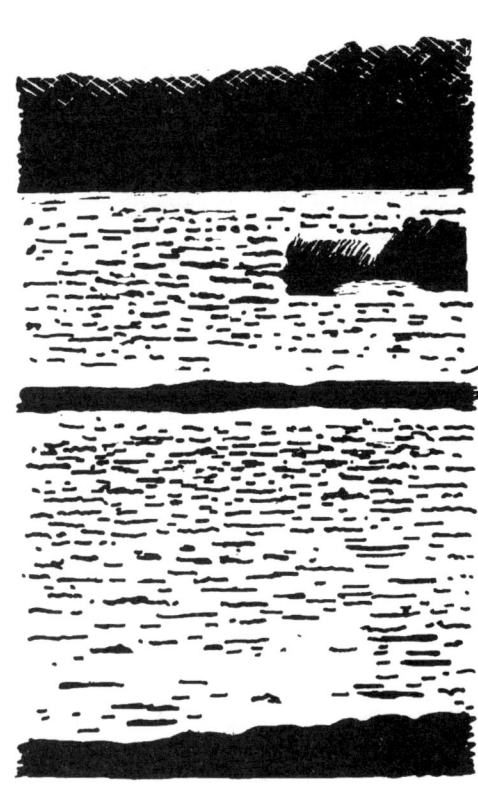

Table 1. TOTAL NUMBERS OF WATERFOWL RECORDED BY WeBS IN GREAT BRITAIN DURING WINTER 1992-93.

Wildfowl at all sites	Sep	Oct	Nov	Dec	Jan	Feb	Mar
Number of sites counted	1,756	1,922	1,911	1,946	2,026	1,993	1,934
Red-throated Diver	107	186	198	642	260	302	423
Black-throated Diver	4	3	23	19	6	14	29
Great Northern Diver	5	9	20	24	24	39	26
Little Grebe	2,600	2,534	2,230	1,877	1,631	1,968	2,072
Great Crested Grebe	9,232	9,580	8,407	7,447	5,425	8,269	8,720
Red-necked Grebe	21	15	18	11	20	34	23
Slavonian Grebe	17	57	79	103	86	163	143
Black-necked Grebe	20	11	7	38	37	36	39
Cormorant	12,175	13,931	11,662	12,198	12,003	11,595	10,776
Grey Heron	2,981	2,939	2,411	2,340	1,984	2,297	2,337
Mute Swan	13,249	13,844	13,688	13,313	13,467	12,561	11,738
Bewick's Swan	0	54	2,668	1,071	6,774	7,016	572
Whooper Swan	7	1,252	3,225	2,698	2,084	2,581	1,793
Bean Goose	0	2	0	2	352	8	39
Pink-footed Goose	12,863	+197,861	+167,512	107,712	71,057	87,830	65,255
European White-fronted Goose	28	106	640	1,529	1,819	1,741	3,088
Greenland White-fronted Goose	0	940	+13,968	391	205	527	+15,221
Lesser White-fronted Goose	1	0	1	0	0	2	1
Greylag Goose: Icelandic*	0	+93,163	+98,144	45,191	28,262	20,380	15,079
Greylag Goose: feral*	15,142	14,773	12,694	13,458	15,084	11,322	10,902
Snow Goose	13	70	67	37	61	76	40
Canada Goose	36,285	35,437	34,545	39,104	34,989	27,788	22,223
Barnacle Goose	190	5,392	10,599	+39,425	5,670	15,580	+29,246
Dark-bellied Brent Goose	2,692	60,134	93,805	97,033	+95,272	95,131	86,819
Light-bellied Brent Goose	466	1,320	+1,790	1,196	1,823	213	25
Red-breasted Goose	1	1	0	0	0	1	0
Egyptian Goose	153	108	77	63	44	51	65
Shelduck	28,241	49,810	56,379	71,914	71,078	67,615	63,582
Mandarin	102	159	141	170	142	119	107
Wigeon	25,120	188,119	188,401	285,868	299,666	170,952	152,342
American Wigeon	0	1	3	1	2	1	2
Gadwall	5,663	7,534	7,762	7,903	7,488	6,968	3,943
Teal	56,452	80,401	92,995	110,048	99,251	78,326	37,763
Mallard	142,260	149,836	158,862	170,954	149,752	105,454	66,068
Pintail	5,248	20,449	15,532	20,993	19,572	13,208	5,481
Garganey	25	2	0	1	3	2	12
Shoveler	7,536	7,785	7,530	7,873	6,186	6,837	5,783
Red-crested Pochard	53	78	134	85	127	74	102
Pochard	14,433	25,155	33,522	33,901	36,490	31,318	12,197
Ferruginous Duck	0	0	2	1	1	1	0
Ring-necked Duck	0	1	4	4	5	0	2
Tufted Duck	37,526	41,103	46,764	52,639	50,105	43,835	35,952
Scaup	289	1,319	1,995	2,291	3,460	3,734	2,384
Eider	27,567	21,416	21,790	14,564	16,809	21,015	25,314
King Eider	0	1	0	0	0	0	0
Long-tailed Duck	5	416	682	969	1,198	1,197	1,028
Common Scoter	1,539	2,178	1,310	1,855	2,275	2,293	1,663
Surf Scoter	0	0	3	1	6	2	2
Velvet Scoter	30	115	205	262	190	265	294
Goldeneye	250	3,049	9,163	12,230	14,844	16,388	13,174
Smew	0	0	14	49	92	108	33
Red-breasted Merganser	2,191	3,343	3,326	3,137	2,778	4,033	3,417
Goosander	906	1,514	1,970	2,410	2,911	2,480	2,171
Ruddy Duck	1,591	2,065	1,912	2,253	2,006	1,997	2,041
Water Rail	73	150	64	142	107	126	146
Moorhen	6,414	7,738	7,412	6,421	7,342	7,216	7,689
Coot	79,460	89,158	95,319	88,416	83,668	56,005	42,121
TOTAL WILDFOWL**	**547,213**	**1,153,747**	**1,222,379**	**1,282,453**	**1,174,051**	**946,848**	**769,235**

notes to Table 1 are provided on page 17.

Waders at estuarine/coastal sites	Nov	Dec	Jan	Feb	Mar
Number of sites counted	146	152	145	153	139
Oystercatcher	291,225	265,463	284,981	205,797	154,847
Avocet	1,776	1,950	1,851	2,137	1,392
Little Ringed Plover	0	0	1	0	0
Ringed Plover	10,495	9,919	8,334	9,464	6,263
Kentish Plover	0	1	1	1	0
Golden Plover	65,072	106,492	55,689	78,203	49,512
Grey Plover	36,967	38,320	38,465	35,513	8,249
Lapwing	186,892	370,455	143,326	216,794	77,463
Knot	289,768	308,468	303,826	164,353	163,398
Sanderling	4,364	6,173	5,008	3,791	4,689
Little Stint	4	3	1	0	0
Curlew Sandpiper	1	0	0	0	0
Purple Sandpiper	1,425	1,421	1,571	1,623	1,698
Dunlin	352,458	448,198	411,368	362,418	307,727
Ruff	117	135	52	149	161
Jack Snipe	48	26	22	32	19
Snipe	2,843	2,049	1,241	1,411	1,289
Woodcock	5	0	2	0	1
Black-tailed Godwit	9,937	8,941	6,394	6,086	8,102
Bar-tailed Godwit	40,764	37,287	40,239	34,083	35,682
Whimbrel	0	4	20	2	8
Curlew	61,495	82,401	62,072	71,194	59,916
Spotted Redshank	69	64	49	53	63
Redshank	70,682	73,000	66,991	62,373	60,680
Greenshank	165	219	136	179	121
Green Sandpiper	38	32	24	46	34
Common Sandpiper	20	14	20	18	12
Turnstone	15,671	14,349	15,895	13,472	13,646
Grey Phalarope	0	0	1	0	0
TOTAL	**1,442,301**	**1,775,384**	**1,447,580**	**1,269,192**	**1,046,807**
Waders at inland sites					
Number of sites counted	1,371	1,366	1,451	1,412	1,410
Oystercatcher	244	436	704	4,828	6,487
Little Ringed Plover	0	0	0	0	4
Ringed Plover	17	11	5	65	175
Golden Plover	16,177	37,684	22,586	20,562	14,641
Grey Plover	3	0	0	2	0
Lapwing	66,288	148,900	127,893	96,965	19,903
Knot	3	0	0	1	0
Sanderling	1	0	0	1	0
Dunlin	508	287	620	1,250	676
Ruff	47	109	225	168	259
Jack Snipe	37	36	36	45	46
Snipe	3,886	3,157	2,063	3,417	2,746
Black-tailed Godwit	3	3	121	4	78
Bar-tailed Godwit	14	1	0	29	0
Curlew	3,526	3,145	3,471	5,915	4,009
Spotted Redshank	1	1	1	1	2
Redshank	507	606	624	813	1,313
Greenshank	4	1	2	3	9
Green Sandpiper	32	39	37	47	64
Common Sandpiper	12	7	5	7	7
Turnstone	27	7	39	8	0
TOTAL	**91,337**	**194,430**	**158,432**	**134,131**	**50,419**
TOTAL WADERS	**1,533,638**	**1,969,814**	**1,606,012**	**1,403,323**	**1,097,226**
TOTAL WATERFOWL[***]	**2,758,428**	**3,254,607**	**2,782,047**	**2,352,468**	**1,868,798**
Kingfisher (all sites)	271	259	206	169	126	151	195

Table 2. TOTAL NUMBERS OF WATERFOWL RECORDED BY WeBS IN NORTHERN IRELAND DURING WINTER 1992-93.

Wildfowl at all sites	Sep	Oct	Nov	Dec	Jan	Feb	Mar
Number of sites counted	110	106	107	99	114	96	100
Red-throated Diver	4	23	40	33	22	14	14
Great Northern Diver	0	0	29	40	18	17	1
Little Grebe	675	789	626	486	571	315	270
Great Crested Grebe	1,750	2,308	708	820	2,081	1,725	1,496
Slavonian Grebe	0	0	51	0	0	4	0
Cormorant	1,849	1,182	1,108	1,243	1,375	1,124	999
Grey Heron	378	347	275	115	147	119	55
Mute Swan	2,290	1,851	2,385	1,747	1,996	1,623	1,241
Bewick's Swan	0	0	46	119	141	222	86
Whooper Swan	4	1,602	2,389	1,114	1,831	1,822	1,133
Pink-footed Goose	4	5	6	0	0	0	0
Greenland White-fronted Goose	0	61	145	0	126	99	122
Greylag Goose*	146	148	383	497	567	772	393
Canada Goose	123	13	346	60	304	468	138
Barnacle Goose	77	76	77	76	75	66	78
Dark-bellied Brent Goose	2	0	0	0	0	0	0
Light-bellied Brent Goose	4,771	10,132	6,236	7,037	6,737	3,618	1,973
Shelduck	135	248	1,183	1,361	2,780	2,368	2,356
Wigeon	3,541	7,194	10,962	4,852	6,475	6,144	4,098
Gadwall	176	289	192	260	314	144	215
Teal	799	2,947	3,272	3,334	3,252	4,437	1,955
Mallard	9,232	8,170	7,473	8,014	6,063	3,737	2,268
Pintail	3	146	212	232	136	140	76
Shoveler	51	112	153	152	114	137	111
Pochard	2,035	10,364	15,332	11,273	23,928	17,062	3,659
Ring-necked Duck	0	0	0	0	0	1	0
Tufted Duck	4,557	14,942	16,502	18,491	19,006	16,643	10,520
Scaup	9	850	982	2,266	2,083	3,567	3,955
Eider	658	663	523	606	568	468	80
Long-tailed Duck	0	3	8	35	16	15	0
Common Scoter	0	117	0	499	0	2	0
Goldeneye	71	2,897	8,458	5,411	14,729	12,601	7,728
Smew	0	0	1	0	0	1	0
Red-breasted Merganser	471	787	825	374	393	468	314
Goosander	0	0	0	0	1	0	0
Ruddy Duck	39	38	0	0	18	0	29
Water Rail	1	5	3	4	2	7	2
Moorhen	373	438	413	241	412	346	282
Coot	7,041	7,935	6,549	10,403	5,111	3,390	1,772
TOTAL WILDFOWL**	**40,509**	**75,869**	**87,082**	**80,762**	**100,791**	**83,178**	**47,065**

Waders at estuarine/coastal sites	Nov	Dec	Jan	Feb	Mar
Number of sites counted	7	8	9	8	7
Oystercatcher	10,942	14,034	14,830	14,992	8,224
Ringed Plover	522	392	851	146	39
Golden Plover	9,006	3,574	10,005	15,034	9,452
Grey Plover	22	75	133	132	37
Lapwing	12,815	16,802	22,395	17,162	10,321
Knot	1,871	292	875	4,487	3,432
Sanderling	2	0	85	0	42
Purple Sandpiper	12	46	87	45	10
Dunlin	5,353	9,263	11,935	9,955	3,107
Ruff	3	0	1	0	0
Jack Snipe	0	0	1	0	0
Snipe	65	35	37	80	22
Black-tailed Godwit	242	152	440	309	314
Bar-tailed Godwit	327	1,423	2,885	1,618	1,122
Curlew	5,030	7,073	5,229	6,440	3,218
Whimbrel	0	0	0	0	3
Spotted Redshank	2	1	0	2	0
Redshank	5,837	6,188	6,143	6,352	3,993
Greenshank	67	65	76	63	33
Turnstone	866	972	2,223	1,014	785
TOTAL	**52,984**	**60,387**	**78,231**	**77,831**	**44,154**
Waders at inland sites					
Number of sites counted	96	86	95	84	89
Golden Plover	3,369	2,677	3,898	4,735	2,545
Lapwing	6,076	7,281	6,709	5,257	218
Common Snipe	19	14	25	9	11
Curlew	418	671	887	853	280
Redshank	9	0	6	6	96
TOTAL	**9,891**	**10,643**	**11,525**	**10,860**	**3,150**
TOTAL WADERS	**62,875**	**71,030**	**89,756**	**86,691**	**47,304**
TOTAL WATERFOWL*	**150,768**	**152,225**	**191,148**	**172,258**	**94,743**
Kingfisher (all sites)	1	0	1	0	0	0	0

* It is not possible to separate the feral from the wild component of this population in Northern Ireland.
** Total wildfowl represents numbers of all divers, grebes, Cormorant, swans, geese, ducks and rails.
*** Total waterfowl represents numbers of all wildfowl (as above), waders and Grey Heron.

MONTHLY FLUCTUATIONS

The vast majority of the wintering population of most wader species is found on estuaries. Coverage of estuaries by WeBS remained at a relatively high and constant level throughout the winter, enabling, for many species, meaningful comparisons of total monthly counts to be made. However, the total numbers of WeBS sites of all habitats counted differ in each month, e.g. a much larger number of inland sites was counted in January to coincide with the International Waterfowl Census. Since wildfowl are more widely distributed across both inland and coastal habitats than waders, changes in monthly count totals given in Tables 1 & 2 may not necessarily reflect true changes in relative abundance during the season. Also, the presentation of data for seven months, which includes the migratory periods for some species, means that there are real fluctuations in total numbers of wildfowl during the period considered in this report.

These fluctuations may be examined by using only counts from sites covered in all seven months (September to March). Once totals from these sites only have been calculated, the number present in each month can be compared directly (expressed as a percentage of the maximum numbers), thus revealing patterns of seasonality for the species considered. This is shown in Tables 3 & 4, for Britain and Northern Ireland separately, giving 1992-93 figures and averages from the last five seasons. Non-migratory, scarce and irregularly counted species are omitted. Caution should be used in interpreting figures for species which only occur in small numbers (see Tables 1 & 2). Thus, numbers tend to fluctuate more widely for many species in Northern Ireland as a result of the smaller numbers of birds involved.

Peak numbers of many of the species considered occurred in December in Britain in 1992-93, with rather fewer peaks than normal in January. Peak abundance for species in Northern Ireland was much more scattered. A smaller proportion than normal of the total of Little Grebes was present during mid-winter in Britain, especially surprising considering the low peak of counted birds. Mid-winter numbers of Great Crested Grebe in Great Britain were similarly low, although the proportion of the peak in Northern Ireland was both markedly lower and higher than expected in different months. There was a noticeable trough in Bewick's Swan numbers in December while Whooper Swan figures were consistently lower than expected in Britain. With the exception of Whooper numbers in October, figures for both species in Northern Ireland were also low throughout the winter. The influx of geese from Holland in March resulted in peak abundance of European White-fronted Geese in that month, more normally recorded in January or February. Numbers of Dark-bellied Brents remained remarkably constant from November to March, presumably since the smaller population resulting from poor breeding success meant that birds wintering on the continent could be accommodated throughout the winter. Gadwall numbers also remained constant and at high levels throughout the winter in Britain, particularly noteworthy in view of the high peak of counted birds. The large fluctuations in numbers for Northern Ireland lacked obvious patterns, although the large deviation from the norm of figures for Pochard coincided with the low national total. Numbers of Goosander in Britain were consistently higher than normal throughout 1992-93.

Table 3. PROPORTIONS IN EACH MONTH OF THE PEAK WINTER POPULATION OF CERTAIN WILDFOWL PRESENT ON 1,403 BRITISH SITES THAT WERE COUNTED IN ALL SEVEN MONTHS OF 1992-93
Bracketed figures give averages for the 1988-89 to 1992-93 period.

	Sep	Oct	Nov	Dec	Jan	Feb	Mar
Little Grebe	100 (100)	88 (89)	74 (82)	62 (74)	52 (70)	65 (63)	72 (63)
Great Crested Grebe	100 (100)	95 (94)	84 (89)	76 (77)	56 (74)	78 (74)	88 (82)
Cormorant	92 (94)	100 (100)	78 (92)	88 (90)	85 (89)	77 (87)	78 (81)
Bewick's Swan	0 (0)	1 (1)	40 (42)	13 (69)	95 (95)	100 (94)	7 (26)
Whooper Swan	<1 (1)	32 (27)	100 (93)	82 (92)	64 (88)	76 (82)	52 (70)
European White-fronted Goose	1 (<1)	3 (2)	21 (12)	51 (52)	61 (85)	58 (88)	100 (43)
Dark-bellied Brent Goose	3 (1)	64 (53)	97 (82)	99 (92)	100 (97)	97 (95)	93 (53)
Shelduck	29 (23)	62 (60)	78 (77)	100 (86)	97 (98)	98 (97)	89 (80)
Wigeon	10 (19)	52 (62)	70 (85)	89 (95)	100 (98)	69 (78)	52 (43)
Gadwall	82 (80)	98 (90)	98 (94)	100 (99)	96 (91)	90 (74)	54 (50)
Teal	50 (51)	73 (68)	85 (81)	100 (98)	89 (88)	74 (76)	35 (32)
Mallard	86 (90)	86 (90)	93 (93)	100 (97)	83 (94)	59 (64)	39 (36)
Pintail	30 (37)	86 (92)	56 (70)	100 (99)	98 (84)	81 (68)	32 (17)
Shoveler	97 (93)	98 (96)	97 (78)	100 (75)	77 (67)	86 (65)	75 (57)
Pochard	39 (34)	71 (64)	92 (86)	93 (88)	100 (100)	86 (86)	34 (31)
Tufted Duck	76 (80)	77 (79)	87 (93)	100 (98)	90 (98)	81 (86)	68 (74)
Goldeneye	2 (2)	22 (15)	60 (57)	77 (82)	97 (94)	100 (99)	90 (90)
Goosander	37 (26)	62 (33)	82 (59)	93 (90)	100 (94)	90 (85)	76 (58)
Coot	89 (92)	98 (97)	100 (95)	92 (91)	86 (84)	59 (64)	46 (43)

Table 4. PROPORTIONS IN EACH MONTH OF THE PEAK WINTER POPULATION OF CERTAIN WILDFOWL PRESENT ON 77 NORTHERN IRELAND SITES THAT WERE COUNTED IN ALL SEVEN MONTHS OF 1992-93.
Bracketed figures give averages for the 1988-89 to 1992-93 period (only 1992-93 data are available for Light-bellied Brent Geese).

	Sep	Oct	Nov	Dec	Jan	Feb	Mar
Little Grebe	82 (89)	100 (90)	78 (87)	63 (74)	71 (58)	37 (48)	30 (33)
Great Crested Grebe	6 (89)	100 (68)	29 (38)	26 (43)	88 (51)	59 (38)	65 (75)
Cormorant	100 (95)	62 (79)	54 (73)	59 (88)	55 (59)	47 (61)	50 (58)
Bewick's Swan	0 (0)	0 (3)	9 (44)	21 (72)	60 (73)	100 (86)	28 (14)
Whooper Swan	<1 (<1)	84 (42)	100 (80)	50 (68)	59 (81)	64 (88)	58 (79)
Light-bellied Brent Goose	47 (-)	100 (-)	62 (-)	67 (-)	63 (-)	33 (-)	19 (-)
Shelduck	5 (4)	10 (9)	45 (40)	49 (82)	100 (87)	81 (82)	90 (71)
Wigeon	35 (24)	70 (75)	100 (82)	45 (59)	55 (73)	52 (57)	40 (33)
Gadwall	58 (79)	96 (82)	64 (83)	86 (80)	100 (88)	48 (64)	71 (59)
Teal	20 (31)	72 (58)	80 (72)	83 (89)	71 (92)	100 (74)	49 (38)
Mallard	100 (98)	92 (85)	80 (69)	89 (68)	61 (57)	39 (39)	26 (22)
Pintail	1 (16)	63 (37)	91 (80)	100 (92)	59 (85)	59 (78)	33 (31)
Shoveler	33 (51)	73 (91)	100 (81)	99 (77)	75 (71)	88 (50)	73 (44)
Pochard	8 (6)	43 (26)	64 (69)	48 (74)	100 (63)	70 (31)	15 (7)
Tufted Duck	24 (20)	80 (50)	87 (77)	100 (89)	100 (90)	85 (70)	56 (47)
Scaup	<1 (<1)	21 (9)	25 (18)	49 (74)	41 (36)	75 (57)	100 (89)
Goldeneye	<1 (2)	20 (9)	57 (59)	37 (79)	100 (73)	85 (73)	53 (55)
Coot	64 (88)	77 (93)	60 (78)	100 (73)	45 (56)	30 (43)	15 (30)

INDICES

As a result of the variable list of sites covered in different seasons, population changes cannot be derived from simply comparing total numbers of waterfowl counted in each year. Consequently, indexing techniques have been developed which allow between year comparisons of populations, even if the true population size is unknown. A new technique has been developed in recent years specifically for waterbird populations, the "Underhill index" (Underhill 1989), which has been adopted for use in this report, replacing the previous "Ogilvie index" used in previous reports. The term "index", when used in this report, refers specifically to the Underhill Index.

Papers fully explaining the new indexing process and its application for major waterfowl populations in the UK will be published shortly (Underhill & Prŷs-Jones in press, Kirby *et al.* in press, a, Prŷs-Jones *et al.* in press). In summary, the index assumes that a count of a species at any site in any month and year can be represented by the combination of a site factor, a month factor and a year factor. The formula derived from this assumption is then used to calculate any missing counts, i.e. when a site was not counted. The year and month factors are set to equal one in a base year and a base month, which varies for each species according to the month of peak abundance. Once the "holes" in the data have thus been filled, the numbers on those sites monitored can be compared between years, and the fluctuations in the population become apparent. Underhill (1989) recommends that, where possible, the index is based on counts from more than one month. The months chosen for each species are given in Tables 5 & 6. The most appropriate grouping of months on which to base the annual index for waders is December, January and February, the period when the wintering population in Britain and Northern Ireland is most stable. However, the peak abundance of different wildfowl species occur in different months according to species, and thus different months and different numbers of months were selected for these species.

Not all species are included in the indexing process however. Notably, many of the goose populations are excluded, partly because their reliance on non-wetland sites requires different count methodologies, but also because regular censusing of the whole of the British populations negates the need for an index to be calculated using the Underhill technique. Thus, the indices for Pink-footed and Greylag Geese have been derived from the highest total count during the October and November census of the population in each year (e.g. Mitchell & Cranswick 1993). Many sea-ducks are also excluded from the indexing process because of the extreme censusing difficulties involved. Waders excluded from the index include those for which large numbers occur away from wetlands, e.g. Lapwing and Golden Plover, and those that are difficult to count accurately using WeBS methods, e.g. Snipe and Jack Snipe. Waterfowl species which only occur in small numbers in Britain and Ireland have also been excluded.

Tables 5 & 6 give index values for waterfowl species in Britain and Northern Ireland, respectively. However, when the indexing model was used to calculate index values for the populations of waders in Northern Ireland, the results were found to be statistically unreliable due to the small number of estuaries contributing to each index value. It was therefore decided to combine the Northern Ireland data with that for Great Britain to produce UK indices for waders and these are given in Table 7.

The size of the species population has been constrained to equal 100 in the base year (1970-71 for wildfowl, 1972-73 for waders). The index values are provided in Tables 5 & 7 as five year summaries extending back to 1966-67 for wildfowl and 1973-74 for waders, representing the first years in which coverage was deemed sufficient for data to be included in the calculation of the index. Due to the shorter run of data available for wildfowl in Northern Ireland, annual index values are provided in Table 6. The annual index figures for the last five years, where calculated, are given in each of the *Species Accounts*.

For all species, the long-term trends in index values indicate significant changes in overall wintering populations. Because short-term fluctuations provide a less rigorous indication of population changes, care should be taken in their interpretation. The indexing technique is still being refined and investigations into the advantages and limitations of the index are ongoing. However, we are confident that this is a major step forward in understanding waterfowl population dynamics.

Wildfowl

The long-term trends in wildfowl populations in Great Britain are discussed in detail in the forthcoming papers. 1992-93 data show declines for many Arctic nesting swans and geese, following a cold, late spring in northern latitudes, with sharp decreases in Bewick's and Whooper Swans being especially noticeable following long-term rises. Index values for many of the duck species, notably Gadwall, Teal, Pintail and Shoveler, also exhibited sharp downturns in 1992-93 following recent increases. Values for Mallard in the last two seasons are noticeably lower than the stable long-term picture, while recent figures for Pochard suggest a continuation of the decrease from the early 1970s. The decline for Gadwall is difficult to explain in view of the peak count (Table 1) and high numbers present throughout the winter (Table 3). Goldeneye numbers continued to rise in 1992-93, whilst Red-breasted Merganser numbers appear to be staging a recovery after last winter's recent decline, although the population remains below the 1980s peak.

This year's report has been taken as an opportunity to apply the new indexing method to those species regularly occurring in the UK, but which were first included in the count scheme only in the 1980s. Thus, annual indices are provided for Little and Great Crested Grebes, Cormorant and Coot, whilst those for Ruddy Duck, excluded by Kirby *et al.* (in press, a), have also been included here. Also, for the first time, populations in Northern Ireland have also been indexed, using the Underhill technique. The methodology used follows that of Kirby *et al.* (in press, a) (see footnote) with the exception that the base year used was 1987-88. Although data for several of the species were collected prior to 1987, data for wildfowl in Northern Ireland are only available from 1985-86. The following winter was selected as the base year to allow for lack of coverage in some areas that inevitably occurred during the first season of coverage. For simplicity, 1987 has been used as the common base year for indexing all of the additional species to Kirby *et al.* (in press, a).

In Northern Ireland, many species appear to have increased in number slightly since the mid and late 1980s. The large increases for Little Grebe and Cormorant probably relate to coverage, since these species were only included in the count scheme in 1985-86 and 1986-87 respectively. Numbers of migrant swans appear to have fallen in recent years, particularly in the case of Bewick's. The indices also suggest a marked reduction in Teal numbers in the last few winters. Gadwall is one of the few species in Northern Ireland to show a sustained increase over the period, mirroring the population rise in Great Britain.

It should be reiterated that comparatively few seasons' data are available for these new species and for Northern Ireland. These indices should be viewed with caution. Many have comparatively large consistency intervals (which provide a measure of confidence in the accuracy of the index, but are not given here due to lack of space). Since the Underhill technique uses data from all available years to calculate index values, future data will refine the index values further.

Footnote: Selection of months was made by calculating monthly index values for all months September to March, and selecting that with the highest index value and any adjacent months with overlapping consistency intervals. Months selected for each species are given in Tables 5 & 6. Data from all available years were used for calculating the index for each species, with the exception of Ruddy Duck, using data from 1966-67 onwards only, as recommended in Kirby *et al.* (in press, a). These were: Little Grebe - 1985-86 onward, Great Crested Grebe - 1982-83, Cormorant - 1986-87, Coot - 1982-83, all species in Northern Ireland - 1986-87. The parameters were used for indexing each species follow Kirby *et al.* (in press, a).

Table 5. LONG-TERM INDICES FOR WINTER WILDFOWL POPULATIONS IN BRITAIN

	Month+	Mean 66-67 to 67-68	Mean 68-69 to 72-73	Mean 73-74 to 77-78	Mean 78-79 to 82-83	Mean 83-84 to 87-88	88-89	89-90	90-91	91-92	92-93
Little Grebe	SO	-	-	-	-	99	289	321	330	313	303
Great Crested Grebe	SON	-	-	-	87	102	109	130	129	124	140
Cormorant	SONDJFM	-	-	-	-	73	138	141	159	144	143
Mute Swan	SONDJFM	115	103	100	105	110	137	146	164	150	150
Bewick's Swan	JF	61	62	89	156	230	189	254	279	297	204
Whooper Swan	ND	102	113	116	129	157	265	251	272	187	185
Pink-footed Goose	O or N	99	96	107	120	173	245	254	266	324	275
Greylag Goose	O or N	88	100	102	131	142	168	129	177	136	152
Canada Goose	S	57	96	124	208	295	422	313	363	408	379
Dark-bellied Brent	DJF	74	86	160	242	297	354	268	350	448	301
Shelduck	JF	82	101	94	126	121	111	118	129	132	117
Wigeon	J	95	91	82	94	125	113	105	108	136	137
Gadwall	SONDJFM	39	89	128	255	388	525	515	551	498	459
Teal	D	70	108	158	242	222	254	315	277	263	230
Mallard	D	90	85	78	84	91	87	81	79	75	75
Pintail	ONDJ	52	91	178	182	198	189	188	163	194	149
Shoveler	SO	100	134	153	166	182	171	186	221	213	160
Pochard	NDJ	116	108	126	99	100	109	104	99	95	91
Tufted Duck	NDJF	80	103	115	113	107	126	99	109	99	102
Goldeneye	F	83	98	112	94	99	102	96	130	122	119
Red-breasted Merganser	ONDJFM	45	95	96	124	127	165	104	131	123	129
Goosander	DJF	64	89	112	156	168	187	117	146	127	121
Ruddy Duck	SONDJFM	25	116	652	2,343	4,695	5,737	6,148	6,103	6,341	4,884
Coot	SONDJ	-	-	-	87	105	111	114	100	98	101

- indicates data are not available for these years
+ the first letter of the months September to March is used to indicate those months used for indexing each species

Table 6. INDICES FOR WINTER WILDFOWL POPULATIONS IN NORTHERN IRELAND

	Month+	86-87	87-88	88-89	89-90	90-91	91-92	92-93
Little Grebe	SON	179	100	372	396	328	354	387
Great Crested Grebe	SONDJFM	76	100	118	92	80	94	117
Cormorant	SOND	10	100	219	273	193	158	226
Mute Swan	SONDJ	88	100	108	120	117	117	116
Bewick's Swan	NDJF	67	100	97	172	189	93	45
Whooper Swan	ONDJFM	77	100	103	80	90	80	85
Light-bellied Brent	SONDJFM	42	100	96	98	118	121	89
Shelduck	DJFM	117	100	173	119	139	115	103
Wigeon	SONDJFM	85	100	124	90	113	103	78
Gadwall	SONDJ	79	100	116	152	122	133	171
Teal	DJ	56	100	100	107	123	92	64
Mallard	SO	138	100	121	141	136	136	117
Pintail	NDJFM	51	100	120	79	124	152	115
Shoveler	SONDJFM	159	100	113	101	92	122	79
Pochard	NDJF	93	100	141	128	131	139	102
Tufted Duck	ONDJFM	52	100	105	111	105	117	109
Goldeneye	DJFM	88	100	83	77	99	107	94
Red-breasted Merganser	SONDJFM	61	100	115	109	105	85	110
Coot	SONDJFM	79	100	99	118	116	108	121

+ the first letter of the months September to March is used to indicate those months used for indexing each species

Waders

Long-term population trends of coastal wintering waders in the UK are presented in the forthcoming paper (Prŷs-Jones et al. in press). In 1992-93, the winter index changed by more than 10% compared to the previous winter for Sanderling, Dunlin and Turnstone. The index for Sanderling dropped by 30% to reach its lowest level for 11 years. A winter index drop of 15% was recorded for both Turnstone (which reached its lowest level for eight years) and for Dunlin (Prŷs-Jones et al. in press). The decline in the Dunlin index is of most concern as it is likely to reflect a real decline in the UK wintering population. The majority of both Sanderling and Turnstone wintering in the UK are present on non-estuarine shores, many of which are not counted for WeBS (Waters in prep.). Low recorded counts for these two species could be the result of birds moving from counted areas to nearby uncounted shores.

Table 7. LONG-TERM INDICES FOR WINTER WADER POPULATIONS IN THE UK

	Month[+]	Mean 73-74 to 77-78	Mean 78-79 to 82-83	Mean 83-84 to 87-88	88-89	89-90	90-91	91-92	92-93
Oystercatcher	DJF	121	134	149	161	157	167	153	165
Ringed Plover	DJF	113	91	100	133	123	113	103	104
Grey Plover	DJF	147	202	308	459	414	462	478	442
Knot	DJF	82	81	96	112	104	106	110	113
Sanderling	DJF	117	108	97	107	87	95	115	81
Dunlin	DJF	108	82	70	83	85	100	97	83
Bar-tailed Godwit	DJF	115	137	143	129	126	152	122	120
Black-tailed Godwit	DJF	87	91	114	167	188	143	168	183
Curlew	DJF	119	111	119	126	135	123	148	151
Redshank	DJF	105	91	98	119	120	101	112	103
Turnstone	DJF	116	110	145	144	152	141	153	134

+ the first letter of the months September to March is used to indicate those months used for indexing each species

SPECIES ACCOUNTS

The following tables rank the principal sites for each species according to average winter maxima calculated from counts received during the last five seasons, 1988-89 to 1992-93. Dashes indicate missing counts and incomplete counts are bracketed. In the first instance, averages were calculated using only complete counts, but if any incomplete counts exceeded this initial average they were then also incorporated and the averages recalculated. Averages enclosed by brackets are based solely on incomplete counts. The month in which the peak 1992-93 count occurred at each site is given in the column labelled "Mth". Yearly index values are provided for the last five seasons where calculated (see *Indices*). A cross (+) denotes counts made during WWT and other goose surveys. Other sources of information are cited accordingly. In wader accounts, non-estuarine coastal sites are identified by an asterisk (*) and inland sites are identified by a hash (#).

International, Great Britain and all-Ireland qualifying levels are given for each species, except where these are unknown (indicated using a question mark "?"), where the population is too small for a meaningful figure to be obtained (indicated using a plus "+"), or where the population is derived from a feral population and (indicated using "introduced"). Note also that, where the 1% level is less than 50 birds, 50 is normally taken as the minimum qualifying level. An asterisk (*) has been used to highlight these instances (see Appendix 1 for a full explanation of national and international qualifying criteria).

The sites included in the tables are, in most cases, those that exceed the appropriate qualifying level for national importance, derived from Kirby (in prep) and Cayford & Waters (in prep) for Great Britain, or all-Ireland importance in Northern Ireland, derived from Whilde (in prep.). The table of sites in each account is divided, where possible, into sites meeting international 1% levels, and those in Great Britain or Northern Ireland meeting either British or all-Ireland 1% levels. For some wader species, not all nationally important sites are given in the tables. In these cases, the names of the sites are listed in the text and the tables flagged accordingly. In cases where the 1% level is very small (less than 30 birds), an arbitrary level has been chosen as the cut off for including sites since data for these species is, in most cases, largely anecdotal. Where all-Ireland criteria have yet to be derived, sites in Northern Ireland have been included according to Great Britain criteria. The locations of sites included in the accounts are given in Appendix 2.

It should be recognised, however, that this report does not provide a definitive statement on the conservation importance of individual sites for waterfowl. The national criteria cut off is chosen for most species to provide a reasonable amount of information in the context of this report only. Thus, for example, many sites of regional importance, or those important because of the assemblage of species present, are not included here. International conservation Directives and Conventions stress the holistic approach needed for successful conservation, and lay great importance on the range of a species, in addition to the conservation of individual key sites.

1992-93 represents only the second season for which several of the wildfowl and associated species have been included in the count scheme. This report is thus still too early for a table of sites regularly supporting large numbers of these species to be drawn up with confidence. Instead, data for sites meeting qualifying criteria in the 1992-93 season only are compared with counts from the 1991-92 season (indicated using the symbol "*cf.*" in the text).

Similarly, counts of waders are now available for many inland wetlands for the last two seasons, 1991-92 and 1992-93. Sizeable populations of Lapwing, Golden Plover, Snipe, Jack Snipe, Ruff and Green Sandpiper were recorded. For these species, count data from inland sites have been included in the following accounts.

RED-THROATED DIVER
Gavia stellata

International importance:		750
Great Britain importance:		50
All-Ireland importance:		10*

GB maximum: 642 Dec
NI maximum: 40 Nov

Trend: not available

The total numbers recorded in Britain in 1992-93 were similar in several respects to those in 1991-92, including a December peak of over 600 birds that greatly exceeded counts in other months. The Northern Ireland totals were much smaller, although the peak was higher than in 1991-92. These figures represent only a small fraction of the 4,300 to 5,400 birds estimated to winter in Great Britain (Danielsen *et al.* 1993). Principal concentrations reflect the easterly bias in the species' distribution in the UK, and counts of more than 50 birds were made at Minsmere (318, December, *cf.* 213 in 1991-92), the Dengie (175, December, *cf.* 150), the Clyde (73, February, *cf.* 9), the Forth (63, November, *cf.* 101) and the North Norfolk Marshes (59, January, *cf.* 26). The principal sites in Northern Ireland were Lough Foyle (34, November, *cf.* 2) and Dundrum Bay (29, December, *cf.* 12), and if suitable seawatching conditions had occurred at both sites in the same month the Northern Ireland totals would undoubtedly have been much larger. The third consecutive winter of comprehensive surveys of the northern half of Cardigan Bay, using aerial, land- and boat-based counts, found a maximum of 390 birds in February (Green & Elliott 1993), although this was fewer than normal for this site (e.g. 963 in 1990-91), partly due to bad weather hampering boat surveys. The majority of birds were located in the coastal strip up to 3 km offshore. A total of 350 birds were recorded by the RSPB/BP counts in the Moray Firth in October (*cf.* 248), with Spey Bay being a favoured site (Evans 1993). It is hoped that future marine SPAs (see *Conservation and Management*) will provide an effective mechanism for the protection of important marine areas such as Cardigan Bay and the Moray Firth.

BLACK-THROATED DIVER
Gavia arctica

International importance:		1,200
Great Britain importance:		7*
All-Ireland importance:		1*

GB maximum: 29 Mar
NI maximum: 0

Trend: not available

Although the maximum count in 1992-93 showed an increase from the previous season, it represents only a small proportion of the British population, which was recently estimated to number 700 (Danielsen *et al.* 1993). The combination of the remote areas favoured by this species, such as southwest Scotland, and the fact that it only occurs in small numbers even in these regions, make for obvious difficulties when censusing. The counts between Turnberry and Dipple (15, March, *cf.* 0 in 1991-92) and in Machrie Bay (12, December, *cf.* 9) are therefore exceptional. The Forth (6, March *cf.* 6) and Kentra Bay (6, November *cf.* 0) were the only other sites to hold more than five birds. The peaks recorded in November and March reflect passage birds, which are perhaps more visible in flocks at this time of year than during mid winter. RSPB/BP counts in the Moray Firth recorded 13 birds in November and December (*cf.* 20), fewer than average owing to the absence of the normal influx in March (Evans 1993).

GREAT NORTHERN DIVER
Gavia immer

International importance:		50
Great Britain importance:		30*
All-Ireland importance:		?

GB maximum:	39	Feb
NI maximum:	40	Dec

Trend: not available

Numbers of Great Northern Divers recorded in Great Britain in 1992-93 were lower than during 1991-92, probably due to counts missing the peak of birds returning north on spring passage. Conversely, counts were higher in Northern Ireland, and would have been higher still if the weather had permitted counts in the same month at the two key sites of Dundrum Bay (40, December, *cf*. 20 in 1991-92) and Lough Foyle (29, November, *cf*. 10), whilst more regular counts at Carlingford Lough (13, January, *cf*. 0) would no doubt have also boosted the total. In Great Britain, the principal concentrations were recorded at Loch Indaal (21, April, *cf*. 22), Machrie Bay (8, February, *cf*. 8) and Lochs Beg and Scridain (6, January, *cf*. 5). Forty birds were present in February in the Moray Firth (*cf*. 17), well above average for this area and particularly noteworthy given the species preference for sites on the west coast (Evans 1993).

LITTLE GREBE
Tachybaptus ruficollis

International importance:		?
Great Britain importance:		30*
All-Ireland importance:		?

GB maximum:	2,600	Sep
NI maximum:	789	Oct

Trend	88-89	89-90	90-91	91-92	92-93
GB	289	321	330	313	303
NI	372	396	328	354	387

The peak count of Little Grebe in Great Britain was well below that of recent seasons, which was as high as 3,500 in 1990-91, and monthly fluctuations showed that numbers declined from the customary September peak more rapidly than usual (Table 3). Although Little Grebe are known to be susceptible to cold winters (Moss & Moss 1993), the relatively mild winter precludes such an explanation in 1992-93. Such an apparently large decline in early winter numbers indicates a poor breeding season in 1992 and the very wet late spring and summer in Britain would have resulted in many nests being flooded, especially on small rivers and canals (Moss & Moss 1993). Numbers in Northern Ireland, however, remained much as normal, although they are prone to fluctuate more markedly from month to month here than in the rest of the UK.

Loughs Neagh and Beg and Strangford Lough remain key sites for Little Grebes. Numbers on the Thames Estuary showed a remarkable increase and this site was exceptional among those in Great Britain in supporting more birds in 1992-93 than expected from its five year average, most supporting fewer than normal numbers. In view of Little Grebes' penchant for quiet, vegetated backwaters, it is perhaps surprising that nine major estuaries feature in Table 8, compared with only two rivers. However, this is partly an effect of the relatively low number of counts received from riverine sites, rather than a reflection of habitat preference. Other sites supporting more than 30 birds in 1992-93 but not listed in the table were Wraysbury Gravel Pits (48, February), Linlithgow Loch (46, September), Portavo Lake (41, October), Kingsbury/Coton Pools (40, September), Blagdon Lake (39, September), Fisherwick/Elford Gravel Pits (39, September), Humber Estuary (39, July), Upper Avon Valley (34, October), North Norfolk Marshes (34, November) and Chatsworth Park Lakes (31, September).

A relatively inconspicuous bird even in winter when it frequents more open waters than during the breeding season, the total number of Little Grebes recorded by WeBS will inevitably be considerably fewer than the true population. Recent estimates suggest 5,000 to 10,000 breeding pairs in Britain (Moss 1993). This would result in a post-breeding population of at least four times that recorded by WeBS in September.

Table 8. LITTLE GREBE: WINTER MAXIMA AT MAIN RESORTS

	88-89	89-90	90-91	91-92	92-93	(Mth)	Average
Great Britain							
Thames Est.	146	104	88	108	182	(Nov)	126
Chew Valley Lake	42	83	100	80	83	(Aug)	78
Deben Est.	45	84	87	69	48	(Dec/Mar)	67
Swale Est.	-	71	108	94	65	(Dec)	65
R. Soar: Leicester	62	67	68	64	43	(Oct)	61
Wash	17	56	112	56	60	(Feb)	60
Medway Est.	39	60	57	53	49	(Dec)	52
R. Test: Full'ton-Stockbr'	-	-	-	39	63	(Sep)	51
Kings Mill Rsr	28	40	46	64	70	(Sep)	50
Morecambe Bay	(0)	86	41	36	30	(Dec/Jan)	48
Rutland Water	73	69	40	27	15	(Sep)	45
Southampton Water	87	50	23	33	25	(Dec)	44
Chichester Hbr	30	52	49	53	36	(Dec)	44
Hanningfield Rsr	85	4	21	46	54	(Aug)	42
Eversley Cross/Yateley GP	16	55	46	37	40	(Mar)	39
The Fleet/Wey	22	26	44	42	46	(Oct)	36
Holme Pierrepont GP	20	4	75	45	95	(Sep)	34
Northern Ireland							
Lo. Neagh/Beg	412	480	324	324	442	(Oct)	396
Strangford Lo.	-	103	122	105	134	(Nov)	116
Upper Lo. Erne	62	57	67	49	27	(Feb)	52

GREAT CRESTED GREBE
Podiceps cristatus

International importance: ?
Great Britain importance: 100
All-Ireland importance: 30*

GB maximum: 9,580 Oct
NI maximum: 2,308 Oct

Trend	88-89	89-90	90-91	91-92	92-93
GB	109	130	129	124	140
NI	118	92	80	94	117

After the apparent setback in 1991-92, the peak counts of Great Crested Grebe in 1992-93 exceeded previous maxima, continuing the trend of increasing numbers since the species was first included in the count scheme. The peak in Great Britain was 8% higher than the previous highest total in 1990-91, and while the count in Northern Ireland was only eight birds more than the previous record count in 1988-90, it was much higher than counts in intervening years. This is borne out by the trends analyses for Great Britain which show a fairly steady rise since the mid-1980s. In Northern Ireland, index values confirm the high numbers in 1992-93 following an obvious low in 1990-91.

Although Loughs Neagh and Beg remains the principal site for Great Crested Grebes in Table 9, peak numbers at this site usually occur in late autumn. This is followed by rapid dispersal so that, in most years, only 200 to 300 birds remain in mid winter. Conversely, numbers at Belfast Lough, which have shown a marked increase over the last five winters, usually exhibit a peak in October or November, with, in recent winters, large numbers being present in other months also.

All UK sites with average maxima of more than 95 birds are given in Table 9. Peak counts vary considerably, from year to year with no discernable pattern. Nevertheless, the 1992-93 peak for Loughs Neagh and Beg, the largest concentration of Great Crested Grebes recorded by the scheme, is exceptional and coincides with a decrease in diving ducks, notably Tufted Duck and Pochard, at the site in 1992-93. Counts on Belfast Lough, the Forth, the Colne, the Mersey and Lough Foyle were also much higher than expected. After several seasons of large counts, numbers at Grafham returned to levels similar to five winters ago, while numbers on the Swale remained low for the third winter in succession. In addition, Loch Ryan (252, September), the Dee Estuary (125, October), Ardleigh Reservoir (115, August), Weirwood Reservoir (112, November), Holme Pierrepont Gravel Pits (107, September), the Thames Estuary (103, March) and Pennington Flash (102, September) held over 100 birds in 1992-93.

Table 9. GREAT CRESTED GREBE: WINTER MAXIMA AT MAIN RESORTS

	88-89	89-90	90-91	91-92	92-93	(Mth)	Average
Great Britain							
Rutland Water	605	544	1,038	878	720	(Sep)	757
Forth Est.	311	849	524	678	923	(Feb)	657
Chew Valley Lake	560	490	440	550	520	(Oct)	512
Grafham Water	179	264	744	522	180	(Nov)	378
Queen Mary Rsr	251	360	526	359	349	(Oct)	369
Cardigan Bay	(190)	-	+385	+376	+322	(Dec)	361
Colne Est.	100	322	214	207	614	(Feb)	291
Morecambe Bay	277	236	229	332	353	(Nov)	285
NE Kent/Thanet	-	-	-	200	339	(Feb)	270
Stour Est.	127	322	200	161	187	(Oct)	199
Pitsford Rsr	202	142	243	243	141	(Dec)	194
Cotswold WP West	136	184	180	200	223	(Mar)	185
Medway Est.	177	254	183	110	135	(Feb)	172
Abberton Rsr	44	303	161	63	247	(Nov)	164
Hanningfield Rsr	130	142	186	233	117	(Nov)	162
Dengie Flats	70	67	34	312	253	(Dec)	147
Blithfield Rsr	69	66	233	166	122	(Nov)	131
Swale Est.	346	160	28	89	27	(Nov)	130
Wraysbury Rsr	-	241	113	87	70	(Oct)	128
Alton Water	-	151	119	93	142	(Dec)	126
Attenborough GP	115	144	142	108	120	(Sep)	126
Mersey	63	112	90	58	277	(Oct)	120
Eyebrook Rsr	30	120	164	112	146	(Sep)	114
S Muskham/N Newark GP	133	-	121	96	79	(Nov)	107
Wraysbury GP	94	108	104	67	156	(Feb)	106
Blackwater Est.	109	85	63	121	122	(Mar)	100
Northern Ireland							
Lo. Neagh/Beg	1,605	1,188	612	753	2,022	(Aug)	1,236
Belfast Lo.	776	886	1,162	1,141	1,771	(Oct)	1,147
Upper Lo. Erne	404	306	137	195	231	(Feb)	255
Carlingford Lo.	106	216	259	279	140	(Dec)	200
Larne Lo.	119	179	88	128	92	(Feb)	121
Lo. Foyle	85	35	60	101	224	(Nov)	101
Dundrum Bay	114	22	68	78	84	(Dec)	73
Strangford Lo.	64	49	67	60	71	(Nov)	62

+ from Green & Elliott (1993)

RED-NECKED GREBE
Podiceps grisegena

International importance: 300
Great Britain importance: 1*
All-Ireland importance: ?

GB maximum: 34 Feb
NI maximum: 0

Trend: not available

Total numbers of Red-necked Grebe recorded in 1992-93 were generally lower than in 1991-92, and the peak count of 34 represents less than one quarter of the British population. The majority of records were of single birds, mainly in southern or eastern Britain, but included records from as far west as Cornwall and Wales. The Forth Estuary again held the largest number of birds (22, February, cf. 32 in 1991-92) and was the only site with more than five birds.

SLAVONIAN GREBE
Podiceps auritus

International importance: 50
Great Britain importance: 4*
All-Ireland importance: ?

GB maximum: 163 Feb
NI maximum: 51 Nov

Trend   not available

Unlike 1991-92, when the number of Slavonian Grebes in Great Britain remained fairly constant across all months, 1992-93 saw a general increase through the winter. The late winter peak of 163 represented an increase of more than 50% on that of the previous season. In Northern Ireland, as in 1991-92, Slavonian Grebes were only recorded in two of the seven winter months, largely as a result of poor weather conditions hampering counts at the key coastal sites where they are found. The high national totals reflect large counts at individual sites, with those at Pagham Harbour (57, February, *cf.* 7 in 1991-92) and Lough Foyle (51, November, *cf.* 8) being particularly impressive, and considerably higher than their respective maxima from the previous season. A further seven sites held 10 or more birds in 1992-93: Loch of Harray (39, December, *cf.* 29), the Forth (32, February, *cf.* 17), the Cromarty Firth (27, February, *cf.* 4), Loch Indaal (19, November, *cf.* 36), Sound of Taransay (12, March, not counted in 1991-92), the Blackwater Estuary (11, March, *cf.* 18) and Poole Harbour (10, December and January, *cf.* 6). RSPB/BP counts on the Moray Firth recorded some 60 birds in December (*cf.* 57).

BLACK-NECKED GREBE
Podiceps nigricollis

International importance: 1,000
Great Britain importance: 1*
All-Ireland importance: ?

GB maximum: 39 Mar
NI maximum: 0

Trend   not available

Apart from the total of just seven birds recorded in November, numbers of Black-necked Grebes in Great Britain in 1992-93 were consistently higher than in 1991-92. The total count from December onwards remained almost constant, with 36 to 39 birds. Birds were recorded at 33 different sites, nearly all in the south or east, with only two in Scotland, one in Wales and, as during last season, none in Northern Ireland. Langstone Harbour was again by far the most important site (28, December, *cf.* 20 in 1991-92), with Loch Ryan (6, March, *cf.* 0) the only other site to support more than five birds.

CORMORANT
Phalacrocorax carbo

International importance: 1,200
Great Britain importance: 130
All-Ireland importance: ?

GB maximum: 13,931 Oct
NI maximum: 1,849 Sep

Trend	88-89	89-90	90-91	91-92	92-93
GB	138	141	159	144	143
NI	219	273	193	158	226

The maximum number of Cormorants counted in Britain and Northern Ireland in 1992-93 each represented the highest counts yet made for the species. This results from sustained increases in both British *(P. c. carbo)* and continental *(P. c. sinensis)* breeding populations, each of which contribute to the British wintering population. British population indices confirm that the winter population is on the increase, and Kirby *et al.* (in press, b), after allowing for incomplete coverage of the species for the WeBS counts, have suggested that the population may well exceed 19,000 birds.

After allowing for differences in the number of sites counted in each month, Cormorants are most numerous on British wetlands in October (Table 3), with a decrease thereafter to levels amounting to 70-80% of the October peak count. In Northern Ireland, the seasonal peak occurs in September and is followed by a sharp decrease to around 50-60% of the peak count. Both patterns presumably arise as a function of winter mortality and movements, the latter including migration overseas as well as movements off the sites covered by the WeBS network.

The principal sites for Cormorants are shown in Table 10, with some 46 UK wetlands represented. Naturally, for a predominantly marine species like the Cormorant, many estuaries rank highly in the list, but the appearance also of many of the larger inland waterbodies is indicative of the species' new found reliance on freshwater wetlands, many of which are stocked with fish for angling purposes. Cormorant numbers fluctuate greatly at individual sites, presumably reflecting the extreme mobility of the species, but particularly striking trends include marked declines on the Medway and Inner Moray Firth, and increases on the Dengie and Exe estuaries. In addition to the sites listed in Table 10, the following wetlands supported more than 130 birds in 1992-93 but failed to qualify for inclusion in the table on the basis of five year average values: Little Paxton Gravel Pits (253, January), Wraysbury Reservoir (246, December), Clwyd Estuary (195, August), Staines Reservoir (183, October), Humber Estuary (181, January), Windermere (174, November), Pagham Harbour (161, November), Chew Valley Lake (160, November), Draycote Water (135, February), Coquet Island (133, February) and Farmoor Reservoirs (130, December).

There is growing conflict between fishery managers and conservationists about Cormorants, particularly with respect to the depredation of inland recreational fisheries. This was discussed in detail at the 3rd International meeting of the Cormorant Research Group in Poland, and resulted in the production of a Position Statement concerning Cormorant Research, Conservation and Management, Gdansk 1993 (Kirby 1993). At this meeting, Kirby & Sellers (in press) provided up-to-date information on status and population trends, highlighting the declines in parts of Scotland despite an overall increase in Britain. Kirby *et al.* (in press, c), reviewing the current position regarding Cormorant conservation and management in Britain, also noted the absence of adequate knowledge of diet in inland situations. They concluded that significant economic damage had not been scientifically demonstrated in Britain and highlighted the inadequacies in government policy with respect to Cormorant control. This does not place Britain in a unique situation since the same could be said for many other countries, but it does emphasise the need for basic research to underpin management plans for the species. The Cormorant Research Group formally affiliated with the International Waterfowl and Wetlands Research Bureau in October 1993, and provides a forum for the discussion of Cormorant issues world-wide.

Table 10. CORMORANT: WINTER MAXIMA AT MAIN RESORTS

	88-89	89-90	90-91	91-92	92-93	(Mth)	Average
Great Britain							
Morecambe Bay	733	1,497	991	1,113	802	(Sep)	1,027
Forth Est.	479	766	962	951	737	(Sep)	779
Medway Est.	415	920	1,216	417	108	(Aug)	615
Inner Clyde	447	663	408	810	565	(Nov)	579
Solway Est.	483	600	492	606	757	(Aug)	588
Loch Leven	270	330	800	390	317	(Dec)	421
Abberton Rsr	-	570	320	380	351	(Oct)	405
Rutland Water	280	250	350	445	532	(Jan)	371
Ranworth/Cockshoot Br.	368	325	329	327	271	(Oct)	324
Queen Mary Rsr	438	315	467	226	124	(Mar)	314
Tees Est.	113	337	480	211	345	(Aug)	297
Grafham Water	32	574	450	350	270	(Nov)	294
Alt Est.	159	334	502	252	143	(Sep)	278
Inner Moray Fth	641	354	112	117	167	(Feb)	278
Dee Est.	290	291	286	201	313	(Sep)	276
Poole Hbr	615	232	417	377	380	(Sep)	273
Colne Est	108	409	169	286	384	(Jan)	271
Ouse Washes	182	533	163	182	248	(Jan)	262
Swale Est.	394	228	263	238	161	(Jul)	257
Blackwater Est.	345	219	208	210	244	(Mar)	245
Wash	294	224	263	206	206	(Nov)	239
Hanningfield Rsr	-	100	374	156	258	(Nov)	222
Carmarthen Bay	-	276	303	151	131	(Sep)	215
Queen Elizabeth II Rsr	99	138	320	430	70	(Sep)	211

	88-89	89-90	90-91	91-92	92-93	(Mth)	Average
Thames Est.	192	255	168	211	204	(Sep)	206
Ribble Est.	242	176	172	161	222	(Oct)	195
Rostherne Mere	109	214	222	159	261	(Feb)	193
Lindisfarne	0	0	720	82	141	(Sep)	189
Barn Elms Rsr	183	160	147	260	118	(Mar)	174
Stour Est.	244	162	124	175	145	(Aug)	170
William Girling Rsr	14	200	232	177	186	(Sep)	162
Dengie Flats	78	51	43	201	401	(Mar)	155
Southampton Water	79	171	171	145	175	(Oct)	148
Blithfield Rsr	141	135	209	144	102	(Sep)	146
Langstone Hbr	166	161	132	149	91	(Aug)	140
Breydon Water	129	180	141	126	122	(Aug)	140
Lo. of Strathbeg	126	128	110	155	175	(Feb)	139
Exe Est.	70	147	83	146	238	(Oct)	137
Alton Water	-	34	208	144	140	(Dec)	132
Northern Ireland							
Lo. Neagh/Beg	591	951	904	446	1,018	(Sep)	782
Belfast Lo.	235	369	284	343	380	(Jan)	322
Strangford Lo.	365	317	119	123	189	(Sep)	223
Outer Ards	379	197	245	146	97	(Sep)	213
Upper Lo. Erne	131	316	192	194	111	(Jan)	189
Carlingford Lo.	-	175	101	174	167	(Dec)	154
Lo. Foyle	65	136	147	188	120	(Nov)	131

GREY HERON
Ardea cinerea

International importance:		4,500
Great Britain importance:		?
All-Ireland importance:		?

GB maximum: 2,981 Sep
NI maximum: 378 Sep

Trend   not available

The number of Grey Herons recorded in Great Britain declined steadily from a September peak to a December minimum before rising again in early spring. This pattern probably results from a combination of mortality, especially in first year birds, and dispersal to more secluded sites not counted for WeBS during mid-winter, with their early return to nesting colonies in spring when they are much more obvious. A similar pattern of decline was observed in Northern Ireland, but instead of a subsequent rise, numbers declined further to just 55 birds in March (Table 1), perhaps reflecting fewer heronries at WeBS sites in Northern Ireland. Total numbers in Great Britain and Northern Ireland were similar to 1991-92, but in view of the 10,300 and 3,600 nests in Britain and Ireland respectively (Marquiss 1993), they represent only a small proportion of the total population. Thirteen sites held 50 or more birds in 1992-93, with maxima occurring in all winter months: Loughs Neagh and Beg (226, August, *cf.* 206 in 1991-92), Deeping St. James Gravel Pits (100, March, *cf.* 152), the Tamar (99, December, not counted in 1991-92), the Thames (90, January, *cf.* 177), Somerset Levels (88, March, *cf.* 97), Strangford Lough (85, November, *cf.* 89), Stockers Lake (70, March, *cf.* 10), Longueville Marsh (68, October, not counted in 1991-92), the Ribble (54, October, *cf.* 9), Poole Harbour (53, September, *cf.* 51), Montrose Basin (52, August, *cf.* 66), the Solway (51, October, *cf.* 37), the River Wye: Chepstow to Tintern (50, February, not counted in 1991-92) and the Dornoch Firth (50, December, not counted in 1991-92).

MUTE SWAN
Cygnus olor

International importance:					1,800
Great Britain importance:					260
All-Ireland importance:					55

GB maximum: 13,844 Oct
NI maximum: 2,385 Nov

Trend	88-89	89-90	90-91	91-92	92-93
GB	137	146	164	150	150
NI	108	120	117	117	116

Total numbers showed the usual pattern of a gradual decline through the winter after an October peak. The decline is too great to be entirely attributable to mortality, and probably reflects the movement, as winter progresses, of birds on to breeding territories where they are missed by WeBS. The maximum count in Britain represented a slight (1.5%) increase on the 1991-92 maximum (13,632) but was still considerably lower than the record 1990-91 total of 15,220. The annual indices show clearly that the population has levelled off since 1990 after a rapid increase in the second half of the 1980s. In Northern Ireland, the maximum count in November was a slight increase on the 1991-92 maximum of 2,278 in October, despite fewer sites having been covered.

Seventeen UK sites have five year average maximum counts indicating national or all-Ireland importance and are listed in Table 11. The maximum count at Loughs Neagh and Beg, the most important site for Mute Swans in the UK, increased slightly after the considerable increase of 1991-92, and the total is getting close to levels which, recorded regularly, would classify the site as internationally important for the species. The Fleet/Wey recorded a slight decline in its maximum count after three years of increase. At the Loch of Harray, Mute Swan numbers crashed to 261 birds, a decline of 54% following the 53% drop in the peak count between 1990-91 and 1991-92. The maximum count at Abberton Reservoir declined by nearly 15% for a second season. Three other sites, the Ouse Washes, the Tweed estuary and the Somerset Levels, recorded spectacular increases in their peak counts to the highest levels in the five most recent years. Of the 18 sites in Table 11, nine recorded declines between 1991-92 and 1992-93 and six increases. All of the increases brought counts to their highest levels in five years, whereas the declines were generally more moderate, with just two sites (Loch of Harray and the Thames estuary) recording their lowest counts for five years. Only one other site, Montrose Basin, recorded a count of more than 200 Mute Swans in 1992-93 (220 in August) in Great Britain, while in Northern Ireland, Ballyroney Lake (59, September) and Belfast Lough (59, December) both held more than 55 birds.

Provisional results of the 1990 breeding season survey showed a population in the spring of that year of 25,750, a 37% increase since the 1983 survey (Delany & Greenwood 1993). The banning of the sale of lead fishing weights was undoubtedly a contributory factor to this increase in England, and probably in Wales, but in Britain as a whole, a succession of mild winters in the late 1980s is also likely to have been important. A paper comparing the findings of the national surveys in 1978, 1983 and 1990 is currently in preparation. The Underhill index technique will be used to impute the number of birds in squares that were not covered in one or more of the years. Totals for each survey year will be calculated and river catchments will be used as the basis for comparison (Greenwood & Delany in press).

Table 11. MUTE SWAN: WINTER MAXIMA AT MAIN RESORTS

	88-89	89-90	90-91	91-92	92-93	(Mth)	Average
Great Britain							
The Fleet/Wey	(571)	891	1,029	1,173	1,126	(Jan)	1,055
Lo. of Harray	655	683	1,205	564	261	(Feb)	674
Abberton Rsr	440	599	635	562	487	(Sep)	545
Ouse Washes	399	544	414	365	615	(Nov)	467
Tweed Est.	240	360	368	370	640	(Aug)	396
Christchurch Hbr	402	538	150	352	210	(Sep)	330
Somerset Levels	271	332	256	252	525	(Jan)	327
Colne Est.	306	316	255	278	325	(Jan)	296
Lo. of Skene	-	175	275	329	375	(Dec)	288
Lo. Bee	-	254	307	-	-		281
Northern Ireland							
Lo. Neagh/Beg	1,120	1,465	1,205	1,601	1,746	(Aug)	1,427
Upper Lo. Erne	336	430	470	520	355	(Jan)	422
Strangford Lo.	212	174	195	114	118	(Nov)	163
Lo. Foyle	90	168	118	102	95	(Sep)	115
Dundrum Bay	164	101	113	79	100	(Nov)	111
Corbet Lo.	156	-	105	100	36	(Nov)	99
Upper Quoile	-	-	-	45	76	(Sep)	61

BEWICK'S SWAN
Cygnus columbianus bewickii

International importance:	170
Great Britain importance:	70
All-Ireland importance:	25*

GB maximum:	7,016	Feb
NI maximum:	222	Feb

Trend	88-89	89-90	90-91	91-92	92-93
GB	189	254	279	297	204
NI	97	172	189	93	45

The number of Bewick's Swans in Britain was well down on recent years. The 1992-93 peak was 23% lower than the 1991-92 maximum (9,118), the highest total yet recorded by WeBS, but also lower than the 1990-91 and 1989-90 maxima (8,444 and 7,905 respectively). Similarly, numbers were well down in Northern Ireland. As in the 1991-92 winter, peak 1992-93 counts were in February, rather later than the December or January peak recorded in earlier years (Table 3). Age counts of birds using the WWT centres indicated an exceptionally poor breeding season with around 5% young recorded at Welney and Martin Mere, and just 3.8% amongst the 398 individuals identified at Slimbridge (Bowler *et al.* 1993).

Numbers at the two principal sites for this species both exceeded their respective five year averages (Table 12) but amongst the top ten sites, only at the Nene Washes was the count a record for the site. Numbers recorded at Loughs Neagh and Beg were particularly low resulting in the return of the Severn Estuary to its position as the fifth largest flock in Britain. The counts at the Somerset Levels, the Wash and the lower Avon Valley were all higher than their respective five year averages, although the last count may have been at the expense of the Mid Avon Valley. Extensive flooding at Walmore Common in December attracted 163 of the Slimbridge birds to the site, a fair return to form following the dry and consequent poor showing of the 1991-92 winter. Other sites holding more than 70 Bewick's Swans were Southampton Water (80, March), the Dee Estuary (77, January) and the River Idle: Bawtry to Misterton (70, February).

The collaborative study of Bewick's Swans breeding in the Nenetski State Game Reserve, North-East European Russia, involving British, Russian, Dutch and Danish scientists, entered its second successive summer. Observations were made of the breeding ecology of the swans from June until August. A total of 60 swans were caught and marked at the Khabuicka study site. A further 310, mostly non-breeding birds, were also marked to the south on the Gulf of Korovinskaia by Dutch, Danish and Russian ringers (Rees *et al.* 1993).

Table 12. BEWICK'S SWAN: WINTER MAXIMA AT MAIN RESORTS

	88-89	89-90	90-91	91-92	92-93	(Mth)	Average
International							
Ouse Washes	3,834	5,984	5,100	5,542	5,169	(Feb)	5,126
Nene Washes	1,137	270	653	1,189	2,543	(Jan)	1,158
Martin Mere/Ribble Est.	639	660	+1,046	+848	+764	(Feb)	791
Breydon Water	698	528	167	394	268	(Mar)	411
Severn Est.	250	+339	340	320	267	(Jan)	303
Lo. Neagh/Beg	246	303	523	232	163	(Feb)	293
Walland Marsh	269	231	-	-	-		250
St Benet's Levels	-	266	182	294	173	(Jan)	229
Avon Valley (Mid)	133	146	296	213	128	(Feb)	183
Berney Marshes	-	187	121	292	113	(Feb)	178
Great Britain							
Somerset Levels	80	222	141	170	209	(Jan)	164
Walmore Common	112	137	164	+97	+163	(Dec)	135
Pulborough Levels	123	78	114	110	66	(Feb)	98
Avon Valley (Lower)	45	121	75	61	129	(Jan)	86
Woodsford Water Meadows	-	-	-	-	79	(Feb)	79
Wash	28	10	16	272	117	(Jan)	76
Northern Ireland							
Lo. Foyle	45	412	195	106	59	(Feb)	163
Boghill Fields	-	54	-	104	31	(Dec)	63
R. Lagan: Moira	-	111	-	11	41	(Dec)	54
Vow Meadows	-	47	-	22	15	(Nov)	28

+ from WWT annual swan reports (e.g. Bowler *et al.* 1992)

WHOOPER SWAN
Cygnus cygnus

				International importance:				170
				Great Britain importance:				55
				All-Ireland importance:				100

			Trend	88-89	89-90	90-91	91-92	92-93
GB maximum:	3,225	Nov	GB	265	251	272	187	185
NI maximum:	2,389	Nov	NI	103	80	90	80	85

The arrival of Whooper Swans at The Wildfowl & Wetlands Trust Centres began in early October, and unusually large numbers were present by mid-month, with peak national totals occurring in November. However, the peak British total remained remarkably low for the second successive season, although the Northern Ireland total had risen considerably. Breeding success amongst Whooper Swans was much higher than for Bewick's, with 9-14% juveniles at WWT reserves (Bowler *et al.* 1993), so the fall in numbers seems likely to be due to the use of other countries within their wintering range. Indices for Great Britain show the population to have fallen slightly in the last two seasons, although numbers remain higher than in the 1960s and 1970s. Index values also indicated smaller numbers in Northern Ireland in recent seasons.

A total of 26 sites support nationally important numbers of Whooper Swans (Table 13). A cold snap in February led to a count of 856 at Welney, the largest concentration ever recorded in England. Numbers on Loch Eye and the Cromarty Firth have returned to more normal levels after the particularly high counts a few seasons ago. The site forms an important staging post for the swans, with peak numbers often recorded in autumn, before the birds disperse south within the British Isles, or in spring, before returning to Iceland. It is thought that the dry summers in 1989 and 1990 resulted in the growth of pondweed which may have provided favourable feeding conditions (J. Bowler, pers. comm.). Consequently, many birds remained longer than usual at Loch Eye in autumn, resulting in the high peaks. Other sites that recorded 56 or more birds in 1992-93 were Upper Nisbet Pond (82, March), Barons Haugh (79, November), Lochhouse Pond (76, December), Merryton Ponds (63, November) and the Lower Derwent Valley (61, January).

The continued study of Whooper Swans at their breeding sites in Iceland by WWT revealed that birds exhibited a high degree of fidelity to their nesting territories. Further, in 1993, one of the 1988-ringed cygnets was found nesting adjacent to the territory in which it had been reared by its parents (Bowler *et al.* 1993).

Table 13. WHOOPER SWAN: WINTER MAXIMA AT MAIN RESORTS

	88-89	89-90	90-919	1-92	92-93	(Mth)	Average
International							
Lo. Neagh/Beg	1,192	1,088	1,110	1,182	883	(Mar)	1,091
Lo. Foyle	1,960	519	988	596	1,166	(Nov)	1,046
Lo. Eye/Cromarty Fth	275	+1,695	+1,115	+340	+389	(Oct)	763
Upper Lo. Erne	582	726	896	889	612	(Feb)	741
Ouse Washes	603	686	578	707	840	(Feb)	683
Lo. of Harray	1,010	817	927	32	19	(Oct)	561
Martin Mere/Ribble Est.	406	572	538	619	530	(Nov)	533
Lo. of Skene	-	406	314	340	425	(Nov)	371
Solway Est.	446	277	96	190	94	(Mar)	221
Lo. of Strathbeg	225	264	129	176	140	(Nov)	187
Great Britain							
Lo. Leven	222	220	180	90	127	(Nov)	168
Loch of Spiggie	74	257	-	165	141	(Oct)	159
R. Tweed: J. Pool-Coldstream	263	135	131	51	139	(Jan)	144
Inner Moray Fth	112	234	87	97	+155	(Dec)	137
Wigtown Bay	75	57	103	80	105	(Mar)	84
Lindisfarne	117	102	82	70	50	(Mar)	84
R. Teviot: Nisbet-Kalemouth	-	-	-	66	82	(Nov)	74
Fairburn Ings	71	81	103	73	67	(Feb)	79
R. Teviot: Kalemouth-Roxburgh	-	-	-	65	80	(Feb)	73
Dinnet Lo.	62	152	75	35	31	(Dec)	71
Glaslyn Marshes	-	-	-	-	64	(Dec)	64
Milldam & Ballfour Mains Pools	-	-	-	66	58	(Nov)	62
Easterloch Uyeasound	61	65	-	56	57	(Nov)	60

+   R.J. Evans in litt

BEAN GOOSE
Anser fabalis

International importance: 800
Great Britain importance: 4*
All-Ireland importance: +*

GB maximum: 352   Jan   Trend   not available
NI maximum: 0

The number of Bean Geese recorded by WeBS returned to a more normal figure after the low of 1991-92. However, in common with other geese, this species often frequents non-wetland areas, and it is often rather hit or miss as to whether birds are present on WeBS sites at the time of the count. This is illustrated by the fact that, with the exception of December and March, fewer than 10 birds were recorded in each month (Table 1). The key site remains the Yare Valley, and 350 birds were recorded at Cantley in December, rather less than were recorded at this site in recent years (Parslow-Otsu 1992). Bean Geese featured in the spring influx of grey geese to eastern England, and 39 birds were recorded away from the two key sites, with 24 at Heigham Holmes being particularly noteworthy. The influx also included many European White-fronts and Barnacle Geese, suggesting that these birds had arrived from The Netherlands.

PINK-FOOTED GOOSE
Anser brachyrhynchus

International importance:	1,900
Great Britain importance:	1,900
All-Ireland importance:	+*

GB maximum: 197,861 Oct
NI maximum: 6 Nov

Trend	88-89	89-90	90-91	91-92	92-93
GB	245	254	266	324	275

The 1992 breeding season was relatively poor for Pink-feet with averages of 9.7% young and 1.67 young per pair present in autumn flocks (Mitchell & Cranswick 1993), the lowest recorded breeding success since 1977. The number csounted in the October and November censuses were 197,000 and 167,000 respectively, the October count being 15% down on the same time in 1991. The counts are believed to be a close estimate of the true population due to good coverage and co-ordination and reasonable counting conditions. However, some observers reported that some birds may have been missed, giving a revised estimate of 200,000-210,000. Over 50% of the October total were found at only five sites: Dupplin Loch, West Water Reservoir, Loch of Stathbeg, Montrose Basin and Loch Leven. No spring census was conducted in 1993 although a total of just over 95,000 birds was counted in February.

Table 14 lists the sites that currently hold 1,900 or more Pink-footed geese, according to the average maxima calculated over the last five seasons. Dupplin Lochs remain the most important site although the peak count of 25,500 in October 1992 was well down on the 57,500 recorded there in 1991 (the largest ever gathering of Pinkfeet recorded in Britain). South-west Lancashire, West Water Reservoir and Loch of Strathbeg continue to support around 30,000 birds, some 10,000 more than the next largest counts. Cameron Reservoir, Montrose Basin, Cowgill Reservoir, Loch Tulybelton, Glenfarg Reservoir, Dun's Dish and Loch Leven all held numbers in 1992-93 in excess of maxima from the previous four years. Sites holding substantially fewer birds than in previous years include Slains Loch, the Wash and Scolt Head, Loch Mahaick and Lake of Menteith. Large numbers of Pink-footed Geese were also recorded at the Tay Estuary (2,800, November), Skinflats (2,596, November), Loch Mullion (2,550, October) and Whitton Loch (2,100, October).

Clearly some sites become more important as the winter progresses with the Wash and Fylde area (Lancashire) supporting more birds in January, and large numbers using Loch of Strathbeg, Solway Estuary, Carsebreck, Castle Loch (Lochmaben) and Wigtown Bay in early spring. A new WWT/SNH project started in October 1993 will explore the distribution of Pinkfeet and Greylags at sites in northern Britain outwith the national census period.

Table 14. PINK-FOOTED GOOSE: WINTER MAXIMA AT MAIN RESORTS

	88-89	89-90	90-91	91-92	92-93	(Mth)	Average
International							
Dupplin Lo.	40,000	31,000	42,000	+57,500	+25,500	(Oct)	39,200
SW Lancashire	30,545	++37,550	++31,805	++38,240	++32,800	(Nov)	34,188
West Water Rsr	40,000	36,250	+24,700	+32,636	+25,000	(Oct)	31,717
Lo. of Strathbeg	30,200	+32,150	+37,100	+23,350	+30,650	(Feb)	30,690
Montrose Basin	+22,000	12,000	15,000	+25,000	+35,000	(Dec)	21,800
Lo. Leven	12,200	+18,000	16,000	+21,880	+23,070	(Oct)	18,230
Slains Lo.	+21,000	+30,300	13,190	-	+4,360	(Oct)	17,213
Hule Moss	5,100	25,735	16,755	+18,500	15,880	(Oct)	16,394
Wash	9,382	19,168	25,330	17,804	8,131	(Jan)	15,963
Solway Est.	+9,006	16,408	17,421	16,210	12,388	(Mar)	14,287
Castle Lo., Lochmaben	2,000	-	16,380	14,000	(3,000)	(Feb)	10,793
Carsebreck/Rhynd Lo.	15,090	11,200	+9,900	+9,250	8,000	(Mar)	10,688
Cameron Rsr	+7,000	+9,500	3,820	+12,270	+15,477	(Nov)	9,613
Aberlady Bay	+7,300	+5,600	17,500	+9,995	+7,000	(Oct)	9,479
Scolt Head	10,180	11,500	8,200	14,000	+3,400	(Nov)	9,456
Fala Flow	3,000	11,920	9,908	+16,410	+4,800	(Oct)	9,208
Fylde/Morecambe Bay	7,900	9,150	+8,240	+9,000	6,100	(Jan)	8,078
Findhorn Bay	9,800	+5,276	-	-	(6)	(Nov)	7,538
Wigtown Bay	+14,000	6,007	6,776	3,810	3,009	(Feb)	6,720
Lo. of Kinnordy	+2,000	8,240	6,980	+6,120	+4,630	(Oct)	5,594
Lo. Tullybelton	+3,050	+3,000	+5,500	+4,500	+5,800	(Oct)	4,370
Cowgill Rsr	3,000	(1,500)	+3,700	+2,800	+6,700	(Nov)	4,050
Lo. Mahaick	+6,531	5,250	4,515	1,471	+800	(Oct)	3,713
Gladhouse Rsr	3,400	5,400	3,200	+2,700	+2,300	(Oct)	3,400
Lo. Eye/Cromarty Fth	+7,000	+1,194	5,500	1,527	800	(Feb)	3,389
Crombie Lo.	6,244	1,391	1,000	+4,250	3,500	(Nov)	2,677
Lake of Menteith	6,000	1,885	+3,600	1,725	80	(Nov)	2,658
Lour	+3,410	+1,800	-	-	-		2,605

++ *from Lancashire Goose Report (e.g. Forshaw 1993).*

EUROPEAN WHITE-FRONTED GOOSE
Anser albifrons albifrons

International importance:	4,500
Great Britain importance:	60
All-Ireland importance:	+*

GB maximum:	3,088	Mar		Trend	not available
NI maximum:	0				

The numbers of European Whitefronts recorded in Britain in 1992-93 were far from normal. Ordinarily, there is a mid- to late-winter peak (Table 4), although the maximum of 1,819 in January was very much smaller than the 6,000 to 7,000 of recent seasons. However, March saw one of the most interesting wildfowl events of the winter, with a large influx of several goose species into the east coast from Norfolk to Humberside. Birds occurred in comparatively large numbers and often in mixed flocks, and European Whitefronts featured strongly amongst these. Indeed, the influx was sufficiently large for the peak number of European Whitefronts to be recorded in March. This is perhaps even more surprising in view of the fact that most over-wintering birds had already returned to the continent by this time. The species complement of the March flocks, which also featured relatively large numbers of Bean and Barnacle Geese, suggests that these birds had originated from The Netherlands.

The effect of the March influx on east coast sites, especially in Norfolk, is evident in Table 15. Additionally, flocks of over 60 birds were found at several sites which do not normally play host to significant numbers of European Whitefronts: the Orwell Estuary (68, March), Sea Bank (Huttoft) Clay Pits (92, February), Kessingland Levels (78, March) and North Warren & Thorpeness Mere (66, March), again, reflecting the easterly bias. The low mid-winter numbers are reflected in low counts from nearly all of this species' favourite haunts, with those at Slimbridge on the Severn Estuary and on the Swale Estuary being particularly notable.

Table 15. EUROPEAN WHITE-FRONTED GOOSE: WINTER MAXIMA AT MAIN RESORTS

	88-89	89-90	90-91	91-92	92-93	(Mth)	Average
Great Britain							
Severn Est.	3,770	3,200	2,600	5,100	1,401	(Jan)	3,214
Swale Est.	2,050	1,660	2,280	1,500	900	(Mar)	1,678
Heigham Holmes	-	-	-	-	350	(Mar)	350
North Norfolk Marshes	376	264	215	163	567	(Mar)	317
Middle Yare Marshes	138	-	295	165	238	(Feb)	209
Thames Est.	300	157	85	178	122	(Mar)	168
Avon Valley (Mid)	245	64	108	221	84	(Feb)	144
Avon Valley (Lower)	71	68	105	172	54	(Jan)	94
Minsmere	180	45	-	108	9	(Feb)	85

GREENLAND WHITE-FRONTED GOOSE
Anser albifrons flavirostris

International importance:	260
Great Britain importance:	140
All-Ireland importance:	140

GB maximum:	15,221	Mar		Trend	not available
NI maximum:	145	Nov			

Greenland Whitefronts have, for 14 years, been the subject of intensive study and monitoring by the Greenland White-fronted Goose Study (GWGS). While many of the birds are counted at some key sites by the WeBS network, coordinated coverage of adjacent areas is undertaken to identify individual flocks and to ensure both that no birds are missed and to avoid double-counting. In particular, complete counts of the principal site of Islay are undertaken with the assistance of Scottish Natural Heritage. These counts are co-ordinated internationally by GWGS and the Irish National Parks and Wildlife Service.

The total counts from the British census' autumn and spring counts are given in Table 1. The March peak was slightly less (1.4%) than the 1991-92 maximum but was higher than those of previous seasons. However, in view of the poor breeding success in 1992, with only 6.2% young recorded from sample flocks (Fox 1993a), it is perhaps surprising that the reduction in total numbers was not larger. Mean brood size, however, was not especially low at three offspring per pair. The March count on Islay was a new record for the island, whilst counts at Rhunahoarine, Tiree and Coll were all much lower than expected (Table 16). Counts at Loch Calder and Scarmclate were also lower, although birds at these sites originate from the Westfield and Heilen flocks respectively, which were presumably less mobile in 1992-93.

The Countryside Council for Wales carried out a thorough survey of the known sites for Greenland Whitefronts in the uplands of central Wales during 1992-93, visiting a number of peatland and other wetland sites which have held geese in the past. Despite their efforts, only one single goose dropping was found and it has to be concluded that the small flock which wintered in the hills above Newtown, which were sporadically encountered up until the early 1980s, must be extinct.

Because of Islay's outstanding international importance as a wintering area for Greenland White-fronted and Barnacle Goose populations, a Goose Management Scheme was introduced in 1992-93 by SNH to encourage sympathetic management of land on the island where these geese occur. Under a voluntary scheme, a system of financial incentives was offered by SNH to support goose use of different areas. Payments were made on the basis of the average numbers of geese, and in return, the recipients agreed to certain specified farming practices which ensured sympathetic management for the birds. This new mechanism has been greatly welcomed, since in the past only farmers who managed areas already designated as being important for geese could receive financial encouragement to manage their holdings for the geese. The Scheme marks a new chapter in goose management on the island, and it will be interesting to assess its long-term effectiveness in reducing conflict on Islay.

Initial results of the WWT study on Islay showed that, although the number of birds wintering on Islay has increased in recent years, the proportion of the population recorded at particular farms was consistent from year to year, reinforcing the view that the geese show a high level of winter-site fidelity. Analysis of the movements of individual birds confirmed that individuals tend to remain within a comparatively small area (with maximum home ranges varying between 42 and 1,500 hectares) and that they make only patchy use of their home ranges. There was substantial variation in reproductive success of the different "sub-populations" both between different parts of Islay and between years, but the reproductive success of birds from particular areas was generally consistent over several years. There was no evidence to suggest that different treatments applied to three experimental fields in summer 1992 (including liming, fertilizing and Juncus-cutting) influenced site selection by the birds during the 1992-93 winter, but is was thought that the changes in farming practice were unlikely to have had a major effect on the vegetation within one season.

Table 16. GREENLAND WHITE-FRONTED GOOSE: WINTER MAXIMA AT MAIN RESORTS[+]

	88-89	89-90	90-91	91-92	92-93	(Mth)	Average
International							
Islay	7,588	8,826	8,857	10,676	11,004	(Mar)	9,390
Machrihanish	907	1,005	1,240	1,023	1,110	(Nov)	1,057
Rhunahaorine	1,116	914	797	1,499	726	(Nov)	1,010
Tiree	728	987	941	1,101	418	(Dec)	835
Coll	647	671	792	621	438	(Dec)	634
Stranraer	393	770	600	438	550	(Feb)	550
Loch Ken	342	550	306	382	323	(Mar)	381
Endrick Mouth	300	300	350	350	250	(Mar)	310
Linne Mhuirich & Lo. na Cille[++]	-	-	-	-	284	(Oct)	284
Great Britain							
Danna/Keills	200	224	245	287	288	(Dec)	249
Westfield Marshes	200	209	180	-	190	(Feb/Apr)	195
Colonsay/Oronsay	165	120	250	210	195	(Dec)	188
Appin/Eriska/Benderloch[++]	76	120	314	270	112	(Dec)	178
Loch Heilen	162	305	160	148	88	(Oct)	173
Bute	-	30	160	250	130	(Dec)	142

[+] including data from GWGS reports (e.g. Fox 1993a)
[++] probably significant numbers of the same birds moving between these sites

GREYLAG GOOSE
Anser anser

Icelandic - International importance:					1,000
Great Britain importance:					1,000
All-Ireland importance:					40*
Hebridean - International importance:					50
Great Britain importance:					50

GB maximum: 110,838 Nov
NI maximum: 772 Feb

Trend	88-89	89-90	90-91	91-92	92-93
GB	168	129	177	136	152

Count conditions and coverage during the 1993 census of Icelandic Greylags were good and produced a total count of 98,144 birds (Table 1). Allowing for missed birds gives a revised population estimate of 100,000, compared with the population estimate in 1991 of 88,000 (Mitchell & Cranswick 1993). Clearly Greylag Geese arrived early in autumn 1992. This was in marked contrast to 1991: between October and November in 1991 the number counted rose from 39,000 to 88,000, whilst in 1992 the October and November counts were quite similar, with an increase of only 5%. In November, over 45% of the total was found at only four sites: Dinnet Lochs, Loch of Skene, Loch Eye and Findhorn Bay - the first two sites alone accounting for one third of the population. In addition to the census counts, a series of counts at Dinnet Lochs recorded over 10,000 birds on 20 occasions with a maximum of 21,650 there on 29 November. Greylag breeding success was also poor in 1992, with only 11.1% young observed in sample flocks, although the mean brood size (2.00) was just below average.

Table 17 lists the sites that currently hold more than 1,000 Greylag Geese, according to the average maxima calculated over the last five seasons. Dinnet Lochs continue to support the bulk of the autumn population with Loch Eye/Cromarty Firth and the Loch of Skene also regularly holding over 10,000 geese, and the Inner Moray Firth, Loch Spynie and Haddo House Loch around half this number. It is interesting to note the gain in importance of the Caithness lochs, Loch of Lintrathen, Drummond Pond, Findhorn Bay, Ballo Reservoir, Lower Bogrotten, Loch of the Lowes and Gadloch as numbers have increased at these sites since 1988-89. Declines in the numbers of Greylags have been recorded at the Inner Moray Firth, Tay/Isla Valley, Loch of Strathbeg, Lindisfarne and Hoselaw Loch over the same period. Other sites holding numbers of Greylag Geese which, if maintained, would be of national significance were: Marlee Loch (1,750 November), Thriepmuir and Harlaw Reservoirs (1,575, November), Loch Fleet (1,570, November), Whiteadder Reservoir (1,480, November), Bemersyde Moss (1,130, October), Loch of Auchlossan (1,050, March), Kilconquhar Loch (1,018, October), Long Loch (1,000, October) and Loch Lomond: Endrick Mouth (1,000, March).

Whilst concern has been raised at the apparent status of the Icelandic Greylag Goose population, the current estimate indicates that the population seems to be stable at about 100,000 individuals rather than declining. Only through continued monitoring of roost sites can the long-term picture be better understood. Co-operation with BASC, hunters and hunting guides is also currently being used to attempt to monitor mortality rates. With recent ringing studies involving capture and subsequent sightings of engraved leg rings, we hope to investigate independent estimates of mortality to complement these.

The population of indigenous Greylags breeding on the Uists is currently being monitored through separate counts and it is hoped to extend these to other Hebridean islands and the north west Scottish mainland. February 1993 produced a post-hunting population of 2,130 birds, maintaining a very gradual increase, although there is considerable hunting on the islands (Mitchell 1993).

Table 17. GREYLAG GOOSE: WINTER MAXIMA AT MAIN RESORTS

	88-89	89-90	90-91	91-92	92-93	(Mth)	Average
International							
Dinnet Lo./R. Dee	+18,000	15,800	+16,000	+18,400	+21,650	(Nov)	17,970
Lo. Eye/Cromarty Fth	+19,259	+11,193	+18,593	+4,659	+16,842	(Oct)	14,109
Lo. of Skene	+8,700	13,305	19,150	+5,298	14,100	(Jan)	12,111
Inner Moray Fth	+12,311	9,271	+8,525	+7,000	1,269	(Jan)	7,675
Lo. Spynie	12,000	3,350	+6,100	+6,600	+7,280	(Oct)	7,066
Haddo House Lo.	5,000	+4,700	+5,900	+6,000	4,200	(Feb)	5,160
Tay/Isla Valley	+6,331	+2,959	+6,262	+4,889	+910	(Oct)	4,270
Caithness Lo.	+2,787	+2,958	+3,064	+4,216	+6,800	(Oct)	3,965
Orkney	2,162	1,817	5,179	4,637	+4,533	(Oct)	3,666
Lo. of Lintrathen	3,050	2,490	+3,600	+3,950	3,900	(Dec)	3,398
Lo. of Strathbeg	6,900	+7,050	925	900	+850	(Oct)	3,325
Drummond Pond	+4,160	+1,800	+3,600	+1,840	+5,050	(Oct)	3,290
Findhorn Bay	(75)	(47)	+3,100	+1,070	+4,900	(Nov)	3,023
Stranraer Lo.	+(1,000)	+2,400	+2,140	+3,300	3,600	(Oct)	2,860
Bute Lochs	x	3,200	+4,200	1,725	2,100	(Dec)	2,806
Lindisfarne	+5,000	+1,700	2,000	2,450	+690	(Nov)	2,368
Holburn Moss	+2,500	3,200	+740	2,750	+2,500	(Nov)	2,338
Fedderate Rsr	+3,300	+2,700	+2,950	+250	-		2,300
Ballo Rsr	-	-	-	+1,420	3,000	(Dec)	2,210
Dornoch Fth	+4,261	2,407	+1,407	927	+1,560	(Dec)	2,112
Hoselaw Lo.	3,600	3,200	+1,270	+1,750	+450	(Oct)	2,054
Lower Bogrotten	-	-	+1,450	+1,200	+3,000	(Nov)	1,883
Lo. Clunie	1,545	375	+4,000	1,260	+2,050	(Oct)	1,846
Lo. of the Lowes	476	(0)	+2,000	+1,142	+3,040	(Oct)	1,665
Corby Lo.	2,000	2,600	1,150	1,400	1,100	(Mar)	1,650
Lo. of Harray	1,446	591	1,420	1,158	2,230	(Feb)	1,369
Lo. Garten	-	-	-	+1,280	+1,057	(Nov)	1,169
Gartmorn Dam	-	(0)	+1,600	+480	+1,200	(Dec)	1,093
Gadloch	800	1,300	+809	807	+1,292	(Nov)	1,002
Northern Ireland							
Strangford Lo.	223	276	546	348	522	(Feb)	383
Lo. Foyle	145	204	452	134	90	(Nov)	205
Temple Water	113	60	163	162	151	(Mar)	130
Larne Lo.	0	27	72	71	34	(Feb)	41

SNOW GOOSE
Anser caerulescens

International importance:	introduced
Great Britain importance:	introduced
All-Ireland importance:	introduced

GB maximum: 76 Feb **Trend** not available
NI maximum: 0

Although some Snow Geese occurring in Britain may originate from the wild populations in North America, the vast majority of records, if not all, undoubtedly relate to birds which originally escaped or were released from collections or zoos. Numbers have increased over the last 25 years, such that while none were recorded during the first breeding Atlas (Sharrock 1976), birds were widely spread throughout Britain at the time of the winter Atlas in the early 1980s (Lack 1986), with breeding recorded both by the New Breeding Atlas (Gibbons et al. 1993) and the recent survey of introduced geese (Delany 1993a). This has lead to the species being placed in categories A and D4 of the British list, indicating that the species has occurred naturally (as a result of vagrancy) and as a feral, but not self-sustaining, population (Vinicombe et al. 1993). The authors urge that the spread of such introduced species be monitored closely to ensure that any subsequent increases do not conflict with the conservation interests of native species.

All monthly totals of Snow Goose in 1992-93 were considerably higher than in 1991-92, probably as a result of observers having become accustomed to recording this species. Most records in 1992-93 were from central England with three sites supporting more than 20 birds: Stratfield Saye (35, February), Blenheim Park Lake (29, October) and Eversley Cross Gravel Pits (24, January). No Snow Geese were recorded at these sites in 1991-92, probably again due to a reluctance to record this species rather than an absence of birds.

CANADA GOOSE
Branta canadensis

International importance: introduced
Great Britain importance: introduced++
All-Ireland importance: introduced

			Trend	88-89	89-90	90-91	91-92	92-93
GB maximum:	39,104	Dec	GB	422	313	363	408	379
NI maximum:	468	Feb						

The peak GB count in 1992-93 was unusual in its timing, occurring in December rather than in September as is more usual (Table 3). Counts of this species in Britain are usually lower at the end of each winter than in September and October, which will be partly a reflection of mortality, chiefly from shooting (Harradine 1991), together with the dispersal of birds to breeding sites in early spring. The December peak was 8% down on that of September 1991-92, and the annual index (Table 5) also reflects this interruption in a long-term increase. In Northern Ireland, the peak was by far the largest since the record count of 550 in February 1986. Three-figure totals were recorded in four other months, and were considerably higher than the 1991-92 counts, suggesting that Canada Geese may be gaining a strong foothold in Northern Ireland.

The largest count in 1992-93 was made at Abberton Reservoir, showing a return to levels recorded in the late 1980s, while numbers at Stratfield Saye declined noticeably compared with recent winters. Spectacular increases in peak counts also occurred on the Stour Estuary and at Kingsbury/Coton Pools, and three sites recorded their first totals of more than 1,000: Dorchester Gravel Pits, Lower Derwent Ings and Port Meadow. Other sites not appearing in the table but holding more than 600 Canada Geese during 1992-93 were Port Meadow (1,000 December), Kingsbury/Coton Pools (859, September), Stour Estuary (803, February), Chew Valley Lake (645, July), Pitsford Reservoir (632, September), Eyebrook Reservoir (609, September) and Colwich Country Park (605, September).

Outstanding data from the 1991 summer survey of introduced geese were received, producing a final total of 63,581 Canada Geese at 1,210 sites in 603 10 km squares (Delany 1993a, Delany in press). This was more than three times the number found during a similar survey conducted in 1976 (Ogilvie 1977), indicating an average rate of increase during the 15 years between the surveys of 8.3% per annum.

Table 18. CANADA GOOSE: WINTER MAXIMA AT MAIN RESORTS

	88-89	89-90	90-91	91-92	92-93	(Mth)	Average
Great Britain							
Stratfield Saye	400	1,350	1,701	2,350	1,090	(Nov)	1,378
Abberton Rsr	1,156	1,240	618	398	1,251	(Sep)	933
Rutland Water	1,102	483	740	1,118	889	(Dec)	866
Bewl Water	1,100	1,000	546	660	943	(Sep)	850
Kedleston Park Lake	1,000	1,080	1,060	570	520	(Oct)	846
Dorchester GP	758	447	767	860	1,075	(Sep)	781
Alde Est.	906	466	538	1,053	932	(Oct)	779
Blithfield Rsr	365	560	896	930	890	(Sep)	728
Lower Derwent Ings	400	-	-	-	1,000	(Nov)	700
Lackford GP	-	-	-	1,000	380	(Dec)	690

++ *since Canada Goose is an introduced species in the UK, a 1% criterion is not used for site designation. A threshold of 600 has been used here as the basis for selecting sites for presentation in this report.*

BARNACLE GOOSE
Branta leucopsis

Greenland - International importance:	320
Great Britain importance:	270
All-Ireland importance:	75
Svalbard - International importance:	120
Great Britain importance:	120

GB maximum: 39,425 Dec
NI maximum: 78 Mar

Trend   not available

About 60% of the Greenland population winter on Islay, where 26,776 were counted in December 1992. The 1992 breeding season was the worst on record, with just 2.6% of the autumn flocks being birds of the year and an average brood size of 1.31. This immediately follows the previous worst-ever season, when there were 4.6% young. The increase in absolute numbers on Islay, combined with the low breeding success, suggest a higher proportion of the population used Islay in 1992-93 (M. Ogilvie, *in litt.*).

Censuses of the whole population are carried out every five years in spring, and a co-ordinated aerial survey in Britain and Ireland was due in March/April 1993. However, despite teams being on standby for two weeks, the weather was never good enough for sufficient time for the survey to be carried out in Scotland. The National Parks and Wildlife Service did, however, carry out the count in Ireland and returned a total of 8,200. This shows that the increase in numbers wintering in Ireland is continuing and, since there has been a considerable increase also in Scotland over the past five years, the population as a whole continues to grow. With 26,500 counted on Islay in March, even if numbers elsewhere in Scotland had not increased, the population was likely for the first time to number more than 40,000 in 1993.

The Islay geese have proved to be an increasing source of conflict with some farmers since the number of birds wintering there have increased from around 5,000 in the early 1960s to a maximum of 30,200 in 1990. Various solutions to the conflict, such as changing shooting seasons, allowing shooting under licence, and making payments to farmers to scare geese from vulnerable crops, have not satisfactorily resolved the problem. On the one hand, conservation organisations press the government to protect the birds as their obligations under the Wildlife and Countryside Act 1981 and the 1979 EC Birds Directive demand, whilst on the other, farmers are unwilling to continue to support an increasing number of geese on their land.

Another solution was tried in 1992-93. The Scottish Office developed a model to assess the impact of geese on farmland on Islay, in terms of yield losses and profits forgone. According to the assessment, payments were made to farmers in relation to the number of geese observed on their land throughout the season according to the fortnightly counts. This scheme rewards farmers for conservation-related benefits rather than for maximising yields. An independent assessment of the Islay situation indicated that, though the financial impact may have been overestimated, the cost of the scheme was probably less than that incurred if land was specially managed on reserves for the same number of birds (Choudhury & Owen 1993). The scheme is being refined and no doubt changes will continue to be made in the light of experience.

The population breeding in Svalbard and wintering on the Solway Firth continues to grow, albeit more slowly than it did in the 1970s and 1980s. Counts in the autumn of 1992 were rather variable, but a co-ordinated survey of the whole of the Solway on 24 October found 12,200, a slight decrease from the highest ever total of 12,700 in 1991-92. Breeding success in recent years has declined considerably (Owen & Black 1991), and 1992 was no exception; there were only 5% young in the autumn flocks (Shimmings *et al.* 1993).

Some of the key UK sites for Barnacle Geese are given in Table 19. However, the logistical problems involved in monitoring areas favoured by the geese, including several of the islands and remote headlands in the north and west of Scotland, means many important sites for these birds do not feature in the table.

Table 19. BARNACLE GOOSE: WINTER MAXIMA AT MAIN RESORTS

	88-89	89-90	90-91	91-92	92-93	(Mth)	Average
International							
Islay	+20,800	+25,297	+30,200	+25,947	+26,776	(Dec)	25,804
Solway Est.	+12,100	+11,700	+12,100	+12,700	+12,200	(Dec)	12,260
Tiree	++894	++581	++1,012	++1,535	++984	(Mar)	1,001
Coll	++218	++343	++275	++670	++3,093	(Dec)	920
Colonsay	++450	++450	++470	++600	++475	(Dec)	489
Danna/Keills	++218	++350	++375	++400	++270	(Dec)	323

++ *Data extracted from the Argyll Bird Report.*

DARK-BELLIED BRENT GOOSE
Branta bernicla bernicla

International importance:	2,500
Great Britain importance:	1,000
All-Ireland importance:	+*

GB maximum:	97,033	Dec		Trend	88-89	89-90	90-91	91-92	92-93
NI maximum:	2	Sep		GB	354	268	350	448	301

Dark-bellied Brents suffered an almost complete breeding failure in 1992 with less than 0.1% young in the population (Cranswick 1993a). Whilst this subspecies is known to exhibit a roughly three-year cyclic fluctuation in breeding success, with one good year, one poor year and one year of variable success (Summers & Underhill 1991), the generally poor conditions that limited the breeding success of nearly all Arctic-nesting swans and geese must have exacerbated the situation. In recent years, WeBS counts have been extended at several of the favourite haunts in January and February, the time of peak abundance in Britain, to locate inland feeding birds and so obtain a full population estimate. The peak count in 1992-93 was, as expected, much lower than the peak of 1991-92, and data received at the end of the 1992-93 season have shown the total to be lower than originally thought (*cf.* Cranswick 1993b). Britain normally plays host to between 40% and 60% of the world population (Salmon & Fox 1991), although the December peak in 1992-93 and the maximum count of 1991-92 represent only 38% and 45% respectively, suggesting that proportionately more birds now remain on the continent.

Surprisingly, comparatively few sites held maxima in 1992-93 that were markedly less than their respective five year average, with numbers on the Wash, the Thames and the Crouch Roach Estuaries showing the largest declines. As expected, only a few sites held more birds. A total of 1,169 birds on the Orwell Estuary in February was the only other count to exceed 1,000 birds.

Table 20. DARK-BELLIED BRENT GOOSE: WINTER MAXIMA AT MAIN RESORTS

	88-89	89-90	90-91	91-92	92-93	(Mth)	Average
International							
Wash	27,612	19,309	21,273	27,742	19,146	(Jan)	23,016
Thames Est.	17,263	12,555	33,109	17,211	15,691	(Oct)	19,166
North Norfolk Marshes	12,711	6,711	11,888	11,128	9,318	(Feb)	10,351
Chichester Hbr	10,473	9,484	9,406	11,582	11,099	(Jan)	10,409
Blackwater Est.	8,363	6,370	9,918	11,445	12,500	(Jan)	9,719
Langstone Hbr	8,050	7,821	6,133	7,860	7,056	(Dec)	7,384
Crouch/Roach Est.	5,333	3,109	8,388	7,978	3,159	(Nov)	5,593
Colne Est.	5,494	3,966	4,924	6,705	6,453	(Jan)	5,508
Hamford Water	3,942	(150)	6,889	4,008	3,677	(Jan)	4,629
Medway Est.	3,093	2,466	5,547	4,484	3,822	(Jan)	3,882
Pagham Hbr	2,965	2,755	3,181	3,669	2,969	(Jan)	3,108
NW Solent	2,400	1,600	3,335	4,868	3,334	(Jan)	3,107
Swale Est.	3,032	1,769	4,823	2,101	1,959	(Jan)	2,737
Humber Est.	(2,000)	1,631	2,733	3,773	2,615	(Dec)	2,688
Portsmouth Hbr	2,062	2,567	2,659	3,580	2,557	(Jan)	2,685
Great Britain							
Fleet/Wey	(0)	850	2,800	4,355	1,982	(Dec)	2,497
Exe Est.	2,795	2,510	2,665	2,020	1,815	(Oct)	2,361
Dengie	2,445	1,900	1,950	2,350	2,320	(Mar)	2,193
Southampton Water	2,486	1,457	1,340	2,752	2,314	(Mar)	2,070
Deben Est.	1,002	2,000	-	3,000	1,555	(Dec)	1,889
Stour Est.	1,784	1,387	1,322	1,980	1,849	(Dec)	1,664
Newtown Est.	1,289	1,117	1,125	1,213	1,664	(Feb)	1,282
Poole Hbr	814	433	1,389	1,711	1,278	(Jan)	1,125
Beaulieu Est.	1,140	740	750	1,110	1,548	(Feb)	1,058

LIGHT-BELLIED BRENT GOOSE
Branta bernicla hrota

Canada/Greenland - International importance:					200
All-Ireland importance:					200
Svalbard - International importance:					40*
Great Britain importance:					25*

GB maximum: 1,823 Jan
NI maximum: 10,132 Oct

Trend	88-89	89-90	90-91	91-92	92-93
NI	96	98	118	121	89

Birds overwintering in Northern Ireland breed in arctic Canada and Greenland and, based on the evidence of data for other species breeding at high latitudes, are likely to have had only moderate or low productivity in 1992. The peak count in Northern Ireland was consequently much lower than in previous years, although the drop from the 15-17,000 of recent winters to around two-thirds of that figure is much greater than might have been expected. Although WeBS data for Northern Ireland are only available from 1985-86 onwards, total numbers of Light-bellied Brents here have been fairly stable, or have perhaps declined slightly. It will be necessary to keep a close eye on numbers in the coming seasons.

Numbers at both Strangford Lough, by far the most important site, and Lough Foyle were lower than usual, with those at the latter site abnormally so. With the exception of Lindisfarne, no other site in Britain regularly holds more than 500 birds. Larne Lough (201, February) was the only other site to support more than 200 birds in 1992-93. Although Light-bellied Brents are recorded in ones and twos quite widely, a series of records of small flocks at Welsh sites, including Swansea Bay (61, September), Inland Sea (27, January) and Carmarthen Bay (22, September), are interesting and probably relate to wandering Irish birds.

Numbers of Svalbard Light-bellied Brents which overwinter on Lindisfarne are monitored weekly throughout the winter for WWT. The number of birds using the site has fallen in recent years, although they increased slightly in 1992-93. Age counts of sample flocks showed these birds to have had a poor breeding season, with only 5.9% young (S. Percival, *in litt.*).

Table 21. LIGHT-BELLIED BRENT GOOSE: WINTER MAXIMA AT MAIN RESORTS

	88-89	89-90	90-91	91-92	92-93	(Mth)	Average
International							
Strangford Lo.	8,478	12,423	13,237	10,359	8,367	(Oct)	10,573
Lo. Foyle	3,700	4,105	6,007	5,395	1,765	(Oct)	4,194
Lindisfarne	3,000	3,000	2,700	1,440	1,865	(Jan)	2,385
Carlingford Lo.	309	259	200	267	243	(Feb)	256
Outer Ards	128	150	418	238	132	(Jan)	213
Dundrum Bay	148	110	183	407	165	(Dec)	203

EGYPTIAN GOOSE
Alopochen aegyptiacus

International importance: introduced
Great Britain importance: introduced

GB maximum: 153 Sep
NI maximum: 0

Trend: not available

The pattern of total numbers of Egyptian Geese recorded in Britain in 1992-93 was very similar to that of 1991-92, with a decline from an autumn peak to a midwinter low, before a slight rise again in spring (Table 1). Although mid winter numbers were very similar in the two seasons, the 1992-93 peak count was very much lower than the 246 in 1991-92. With its apparent reluctance to spread from its original stronghold in Norfolk, the vast majority of records came from this and adjacent counties, although birds were recorded as far afield as Devon and Greater Manchester. The following sites held more than 10 birds in 1992-93: Sennow Park Lake (58, August, not counted in 1991-92), St Benet's Levels (54, October, *cf.* 29), Pentney Gravel Pits (32, September, *cf.* 88), the North Norfolk Marshes (19, November, *cf.* 71), Rutland Water (16, March, *cf.* 9), Blickling Lake (14, September, *cf.* 23) and River Wensum: Fakenham (12, September, *cf.* 7).

SHELDUCK
Tadorna tadorna

			International importance:				2,500
			Great Britain importance:				750
			All-Ireland importance:				70

			Trend	88-89	89-90	90-91	91-92	92-93
GB maximum:	71,914	Dec	GB	111	118	129	132	117
NI maximum:	2,780	Jan	NI	173	119	139	115	103

Shelducks are highly mobile in winter and the pattern of occurrence in the UK showed the typical strong build-up in numbers from September to a mid-winter peak followed by a slight decline to March (Tables 3 & 4). The December maximum in Britain showed a 15% reduction from the record peak of 84,017 in 1991-92. In Northern Ireland, the build-up and decline were, typically, a little later than in Britain, with the peak some 300 up on 1991-92, but still 700 fewer than the 1990-91 peak. This seasonal pattern reflects the gradual return of British and Irish breeding birds from the moulting grounds on the German Wadden Sea, with the addition of birds moving through from other parts of northwest Europe. The annual index (Table 5) shows that the number of Shelducks counted at British sites in winter is increasing overall, but that fluctuations in this growth are nothing unusual.

Twelve sites have recorded average maximum counts over the last five winters which make them internationally important for Shelduck; these appear at the top of Table 22, together with the 24 sites which are of British and all-Irish importance for the species. All these sites are large estuaries, pre-eminent among which is the Wash, which has held two-and-a-half to three times as many Shelducks as any other UK site in each of the last five winters. The peak count at the Wash was, however, 9% lower than in 1991-92, and seventeen other sites appearing on the table recorded declines in their winter maximum counts. The greatest of these declines occurred at the Burry Inlet, and a little further west, the Cleddau also suffered a considerable drop in numbers. The greatest increase occurred on the Swale, where the peak count was the highest in five years. Three further sites recorded more than 750 Shelducks, namely the Crouch/Roach Estuary (1,046, December), the Tamar Estuary (853, March) and the Blyth Estuary in Suffolk (762, March).

The 1992 breeding survey was extended by means of a "mop-up" survey in the summer of 1993 which resulted in almost complete coverage of Britain. Just over 1,000 sites in 663 10 km squares were covered. A provisional analysis of the survey data has revealed totals of 44,700 adult Shelduck in Britain between late April and mid May, with almost half of the population at this time having been recorded in pairs. This total is just 500 more than that found by extrapolation from tetrad counts between 1988-91 for the New Breeding Atlas, a remarkable vindication of the methods used in both surveys. Analysis of the 1992-93 survey data will allow the relative importance of individual sites and regions for the Shelduck in summer to be assessed in considerable detail. Preliminary analysis of data collected on brood sizes produced a provisional total of 12,800 juveniles and suggests geographical variation in brood sizes: for example, in Scotland, northern Cumbria and Northumberland in 1992, modal brood size was three, whereas in the rest of England and Wales it was five.

Table 22. SHELDUCK: WINTER MAXIMA AT MAIN RESORTS

	88-89	89-90	90-91	91-92	92-93	(Mth)	Average
International							
Wash	15,613	19,460	16,275	20,194	18,465	(Dec)	18,001
Dee Est.	4,896	6,924	6,617	6,487	6,893	(Oct)	6,364
Medway Est.	2,985	5,092	7,963	6,068	5,585	(Jan)	5,539
Morecambe Bay	3,345	5,208	6,143	6,972	5,178	(Oct)	5,369
Mersey Est.	2,602	4,040	5,757	7,946	4,414	(Oct)	4,952
Humber Est.	4,681	4,245	5,856	4,680	3,971	(Oct)	4,687
Forth Est.	++2,400	2,670	++4,025	++4,420	4,414	(Sep)	3,586
Ribble Est.	3,534	3,162	3,113	4,849	3,040	(Jul)	3,540
Severn Est.	2,819	+3,332	3,598	3,644	2,560	(Feb)	3,191
Blackwater Est.	2,000	2,599	3,398	2,960	3,356	(Jan)	2,863
Thames Est.	2,351	3,137	2,535	2,515	3,097	(Feb)	2,727
Poole Hbr	2,230	2,179	3,451	2,382	2,769	(Mar)	2,602
Great Britain							
Chichester Hbr	2,514	2,717	2,321	1,863	1,750	(Feb)	2,433
Swale Est.	1,233	1,529	2,609	2,286	3,692	(Jan)	2,269
Stour Est.	1,578	1,752	1,589	2,822	2,272	(Jan)	2,003
Solway Est.	2,074	1,851	1,551	2,090	2,294	(Jul)	1,972
Colne Est.	867	1,164	1,191	1,942	2,337	(Mar)	1,500
Tees Est.	1,365	1,637	1,404	1,154	1,138	(Jan)	1,340
Eden Est.	1,221	1,121	1,176	1,590	1,133	(Jan)	1,248
Burry Est.	1,655	1,488	1,360	1,504	218	(Jan)	1,245
Alde Est.	1,315	1,128	1,474	1,154	1,035	(Nov)	1,221
Hamford Water	1,224	(98)	1,009	1,202	1,423	(Feb)	1,214
N. Norfolk Marshes	819	1,861	1,161	1,064	1,084	(Dec)	1,198
Orwell Est.	1,249	1,104	-	1,026	1,411	(Jan)	1,197
Deben Est.	1,409	1,120	-	875	779	(Feb)	1,046
Langstone Hbr	1,040	1,203	1,017	740	1,017	(Jan)	1,003
Lindisfarne	780	800	1,140	1,065	825	(Jan)	922
Cleddau Est.	950	969	1,038	923	613	(Jan)	899
Duddon Est.	1,046	852	942	865	770	(Jan)	895
Northern Ireland							
Strangford Lo.	3,973	1,867	2,311	1,950	1,755	(Jan)	2,371
Belfast Lo.	170	465	427	467	287	(Jan)	363
Lo. Foyle	240	315	421	379	179	(Jan)	307
Larne Lo.	241	240	274	225	248	(Mar)	246
Carlingford Lo.	188	301	257	176	243	(Feb)	233
Lo. Neagh/Beg	211	215	142	159	189	(Mar)	183
Dundrum Bay	113	76	109	90	143	(Mar)	106

+ Counts of breeders and non-breeders from Jones (1989).
++ Counts in August of moulting birds (D.M. Bryant, in litt.).

MANDARIN
Aix galericulata

GB maximum:	170	Dec
NI maximum:	0	

International importance: introduced
Great Britain importance: introduced
Trend not available

Numbers of Mandarin recorded by WeBS in Britain during 1992-93 showed a rise to a midwinter peak, before falling again in spring (Table 1), a similar pattern to that of 1991-92 although the peak was then more pronounced, with a maximum of 240 birds. Unlike last season, no birds were recorded in Northern Ireland. Records were concentrated in the home counties and central England, but birds were also recorded in Scotland and Wales. A total of 11 sites held 10 or more birds, with Woburn Park Lakes (44, October, *cf.* 53 in 1991-92), Cuttmill Ponds (42, January, *cf.* 40), Swanbourne Lake (31, December, *cf.* 11), Busbridge Lakes (23, February, *cf.* 31), Witley Park (18, November, *cf.* 25), Frenchess Pond (16, January, not counted in 1991-92) and Virginia Water (15, October, *cf.* 37) being prominent amongst these.

WIGEON
Anas penelope

GB maximum:	299,666	Jan
NI maximum:	10,962	Nov

International importance: 7,500
Great Britain importance: 2,800
All-Ireland importance: 1,250

Trend	88-89	89-90	90-91	91-92	92-93
GB	113	105	108	136	137
NI	124	90	113	103	78

Numbers of Wigeon in Britain again rose steadily to a January peak, which, although less than the 1991-92 record count, remained very high, as borne out by the indices, and represents the largest count of any wildfowl species by WeBS in 1992-93. Far fewer birds winter in Northern Ireland, although the peak total was rather small compared with previous years. However, the Northern Ireland maximum, usually in October or November, is often very much larger than the totals in other months (Table 4), which, in 1992-93, were comparable with recent seasons.

Sites supporting over 2,800 Wigeon are given in Table 23, including 10 sites of international importance. Numbers on the Ribble Estuary returned to more normal levels after the record count last season, although the December peak again represents the largest count of any wildfowl species at a single site in 1992-93. A number of sites held rather fewer birds than might be expected, notably Lough Foyle, which supports the majority of birds in Northern Ireland, the Medway and Montrose Basin, whilst counts at the Loch of Harray remained low for the second season running. The low count on the Cromarty Firth appears to have been compensated for by higher counts on the adjacent Dornoch and Inner Moray Firths. Large numbers were recorded on the Swale, with those on the Mersey and the Humber continuing the general increase in importance of these sites in recent years. Large counts at Loch of Skene and the Stour in 1992-93 elevated these sites to national importance for Wigeon. The count on the Somerset Levels was particularly impressive, especially since nearly 7,000 of these were found at West Sedgemoor alone, in the company of over 10,000 Teal and 23,000 Lapwing. Other sites holding in excess of 2,800 Wigeon were Breydon Water (3,800, December), Stour Estuary (3,757, January), Berney Marshes (3,250, January), Pulborough Levels (3,176, January) and the Blackwater (2,928, January).

Table 23. WIGEON: WINTER MAXIMA AT MAIN RESORTS

	88-89	89-90	90-91	91-92	92-93	(Mth)	Average
International							
Ribble Est.	41,809	43,541	59,187	88,612	48,441	(Dec)	56,318
Ouse Washes	30,968	53,615	24,715	37,064	28,879	(Jan)	35,048
Lo. Foyle	22,000	7,797	15,584	16,622	5,869	(Nov)	13,574
Dornoch Fth	10,299	13,861	10,251	17,637	15,091	(Oct)	13,428
Lindisfarne	28,000	7,500	9,040	9,580	5,845	(Oct)	11,993
Martin Mere	18,000	8,000	2,200	16,630	11,220	(Jan)	11,210
Cromarty Fth	8,158	9,686	10,977	14,878	7,299	(Dec)	10,200
N. Norfolk Marshes	6,580	6,825	12,779	14,898	9,881	(Jan)	10,193
Swale Est.	6,801	8,625	11,673	9,731	12,422	(Mar)	9,850
Inner Moray Fth	6,702	7,775	7,430	8,058	9,999	(Dec)	7,993
Great Britain							
Mersey	4,630	4,000	6,590	11,500	9,235	(Oct)	7,191
Lower Derwent Valley	6,150	-	-	-	7,500	(Mar)	6,825
Fleet/Wey	(0)	4,798	6,181	8,245	5,804	(Nov)	6,257
Buckenham Marshes	7,050	-	7,000	5,410	5,358	(Feb)	6,204
Lo. of Harray	7,455	5,551	9,200	3,780	3,285	(Nov)	5,854
Medway Est.	6,276	4,971	7,610	5,741	2,866	(Jan)	5,493
Morecambe Bay	4,404	4,667	4,084	6,944	5,832	(Nov)	5,186
Montrose Basin	5,781	4,852	5,355	5,456	3,555	(Nov)	5,000
Rutland Water	5,142	3,261	4,380	4,270	3,877	(Feb)	4,186
Severn Est.	4,557	4,017	3,935	3,910	3,838	(Dec)	4,051
Nene Washes	3,260	7,958	2,057	2,050	4,708	(Jan)	4,007
Dyfi Est.	4,537	4,422	3,334	4,003	3,689	(Jan)	3,997
Somerset Levels	(6)	798	1,108	3,480	10,253	(Jan)	3,910
Wash	2,495	1,977	4,009	5,441	3,203	(Sep)	3,425
Humber Est.	2,465	2,437	3,038	4,767	4,349	(Oct)	3,411
Alde Est.	2,821	3,216	3,843	3,293	2,893	(Dec)	3,213
Cleddau Est.	2,489	3,675	2,951	4,020	2,461	(Dec)	3,119
Dee Est.	2,180	827	4,307	5,078	3,155	(Nov)	3,109
Lo. of Skene	-	1,393	3,800	2,018	4,098	(Dec)	2,827
Northern Ireland							
Lo. Neagh/Beg	2,521	2,607	5,949	3,203	2,849	(Feb)	3,236
Strangford Lo.	1,382	1,538	1,369	1,630	1,900	(Nov)	1,564

GADWALL
Anas strepera

NW European importance: 250
Great Britain importance: 80
All-Ireland importance: +*

GB maximum: 7,903 Dec
NI maximum: 314 Jan

Trend	88-89	89-90	90-91	91-92	92-93
GB	525	515	551	498	459
NI	116	152	122	133	171

The peak count of Gadwall recorded in Britain reached an all-time high in 1992-93, although only slightly larger than in recent seasons. Index values suggest the rapid increase in the size of the population during the 1980s has slowed in recent years (Table 5), with a sharp fall in 1992-93 despite the increase in the peak count. The Northern Ireland population remains small, with numbers rising slightly in 1992-93, though monthly fluctuations are very erratic (Table 4). Over 90% of the peak British total were present from October to February (Table 3), possibly as a result of the mild conditions in the latter part of the winter.

Table 24 lists all sites which have held 80 or more Gadwall on average over the past five seasons. Seven of the top 10 sites held below average numbers, with remarkably low counts at Rutland Water and Gunton Park Lake. However, numbers continue to rise on many other sites, notably on the Mid Avon Valley, Wraysbury Gravel Pits and Hardley Flood. The numbers at Loughs Neagh and Beg and at Strangford Lough are particularly noteworthy in view of the small total numbers recorded in Northern Ireland. Other sites recording 80 or more birds in 1992-93 were South Iver Gravel Pits (121, September), Old Slade Complex (121, September), Twyford Gravel Pits (98, October), Wotton Underwood Lakes (90, February), Hollowell Reservoir (89, September), Stanford Reservoir (89, January), Allington Reservoir (88, December), Thoresby Lake (88, September), Church Wilne Reservoir (87, December), Crouch/Roach Estuary (86, January), Meadow Lane Gravel Pits (81, January), Rye Harbour and Pett Level (81, November), Somerset Levels (80, February) and Sennowe Park Lake (80, October).

The rising numbers and spread of Gadwall across southern England have been well documented, and may be due, at least in part, to the species' feeding habits. Though essentially a dabbling duck, it may often be found on deeper waters associating with Coot. Amat & Soriguer (1984) noted Gadwall kleptoparasitising diving Coot by feeding on the submerged vegetation the Coot brought to the surface, hence opening up a whole new food supply which was previously unavailable. This may have made many waterbodies, which would otherwise be unsuitable, available to Gadwall, and facilitated their rapid spread.

Table 24. GADWALL: WINTER MAXIMA AT MAIN RESORTS

	88-89	89-90	90-91	91-92	92-93	(Mth)	Average
International							
Rutland Water	1,805	1,606	1,323	1,369	501	(Sep)	1,321
Avon Valley (Mid)	333	366	493	719	1,051	(Jan)	592
Abberton Rsr	784	846	402	218	358	(Sep)	522
Gunton Park Lake	461	496	325	450	38	(Sep)	354
Severn Est.	290	384	345	296	332	(Jan)	329
Ouse Washes	229	379	352	249	417	(Feb)	325
Thorpe Water Park	-	-	-	285	250	(Nov)	268
Chew Valley Lake	160	351	190	425	190	(Aug)	263
Great Britain							
Cheshunt GP	200	335	290	205	130	(Jan)	232
Lackford GP	-	-	-	217	161	(Dec)	189
Lo. Leven	154	163	258	120	200	(Sep)	179
Buckden/Stirtloe GPs	86	272	-	209	144	(Nov)	178
Stanford Training Area	110	133	141	242	215	(Oct)	168
Hickling Broad	81	188	312	137	113	(Feb)	166
Seaton GP	-	-	-	-	163	(Nov)	163
Nene Washes	54	303	87	62	239	(Feb)	149
Wraysbury GP	14	61	116	116	404	(Dec)	142
Cotswold WP West	108	156	166	107	170	(Nov)	141
Thrapston GP	181	106	123	168	97	(Sep)	135
North Norfolk Coast	99	197	123	123	133	(Dec)	135
Amwell GP	78	66	209	93	215	(Oct)	132
Hardley Flood	16	20	73	128	373	(Sep)	122
Ware GP	-	-	-	-	122	(Oct)	122
Tattershall GPs	-	174	43	62	160	(Nov)	110
Thames Est.	83	183	37	83	133	(Feb)	104
Cheddar Rsr	41	153	112	105	61	(Dec)	94
Summerleaze GP	47	161	80	43	130	(Feb)	92
Minsmere	65	128	-	90	82	(Oct)	91
Rye Meads Sewage Farm	-	40	118	92	101	(Dec)	88
Swillington Ings	37	124	-	100	82	(Oct)	86
Bucklands Pond	24	174	27	33	154	(Nov)	82
Northern Ireland							
Lo. Neagh/Beg	118	120	88	133	158	(Oct)	123
Strangford Lo.	75	144	94	106	114	(Jan)	107

TEAL
Anas crecca

				International importance:				4,000
				Great Britain importance:				1,400
				All-Ireland importance:				650

GB maximum:	110,048	Dec
NI maximum:	4,437	Feb

Trend	88-89	89-90	90-91	91-92	92-93
GB	254	315	277	263	230
NI	100	107	123	92	64

Peak numbers of Teal in Britain in 1992-93 fell by 6% compared with 1991-92, having fallen by almost 20% the previous winter. Similarly, the peak count in Northern Ireland showed a reduction of around 9% compared with 1991-92. Teal numbers in Britain have shown a general increase during the last 25 seasons (Table 5, Gilburn & Kirby 1992), mirroring the trend observed in the northwest European population. The rapid increase in the late 1980s to record levels has been followed by declines in the last two seasons which have placed the trend back on a more moderate rate of increase. In Britain, monthly fluctuations were similar to those of previous seasons, with high numbers present in November through January with a peak count in December (Table 3). Thereafter, numbers declined gradually, with a sharp decrease in numbers by the March count. In Northern Ireland, the numbers were relatively high and stable from October onwards with a sudden jump to the peak count in February before falling off steeply in March (Table 4).

1992-93 saw a consistent reduction in numbers across virtually all of the main resorts for Teal (Table 25), with declines at over 60% of the sites listed. This is in marked contrast to 1991-92 when, despite the low national totals, many sites held record counts; half of the declines shown in Table 25 were between 30% and 50% lower than 1991-92. Of the 10 sites which held numbers in excess of their five year averages, the Somerset Levels was exceptional, with the majority of birds being found at the RSPB reserve of West Sedgemoor. This coincided with huge numbers of Wigeon, Lapwing and Golden Plover at this site also. Similarly, the large counts made on the Lower Derwent Ings and at Pulborough Levels highlight the benefits to be gained from positive management of this habitat (Green & Robins 1993). Two further sites held 1,400 or more Teal during the 1992-93 count period: Pagham Harbour (1,404, November) and Mid Avon Valley (1,304, January).

Table 25. TEAL: WINTER MAXIMA AT MAIN RESORTS

	88-89	89-90	90-91	91-92	92-93	(Mth)	Average
International							
Mersey Est.	9,670	12,300	10,375	13,450	12,020	(Dec)	11,563
Dee Est.	4,670	9,825	4,824	10,715	6,194	(Jan)	7,246
Ribble Est.	6,417	1,709	9,078	9,500	6,813	(Nov)	6,703
Somerset Levels	(500)	4,514	2,808	1,908	11,330	(Feb)	5,174
Abberton Rsr	1,850	4,225	11,483	4,245	2,321	(Sep)	4,825
Ouse Washes	3,870	4,920	5,225	5,157	2,085	(Feb)	4,251
Great Britain							
Hamford Water	1,975	(677)	7,211	4,048	2,184	(Oct)	3,854
Lower Derwent Ings	3,300	-	-	-	4,132	(Feb)	3,716
N. Norfolk Marshes	2,337	5,538	3,223	2,740	2,779	(Dec)	3,323
Woolston Eyes	3,500	4,000	4,500	1,500	1,550	(Dec)	3,010
Martin Mere	4,300	2,600	1,900	2,600	3,470	(Oct)	2,974
Cleddau Est.	3,243	2,586	3,148	2,188	1,671	(Jan)	2,567
Medway Est.	3,523	1,827	2,992	2,360	1,619	(Dec)	2,464
Lo. Leven	1,400	3,270	3,614	1,873	1,850	(Sep)	2,401
Morecambe Bay	2,349	2,410	1,421	3,036	2,349	(Jan)	2,313
Thames Est.	1,996	3,342	3,407	1,627	1,058	(Jan)	2,286
Severn Est.	1,253	3,402	1,820	2,711	1,986	(Dec)	2,234
Inner Moray Fth	1,780	2,087	2,800	2,225	1,957	(Jan)	2,170
Blackwater Est.	3,897	1,779	1,341	2,002	1,558	(Nov)	2,115
Chew Valley Lake	725	1,115	5,500	1,300	1,765	(Sep)	2,081
Alde Est.	2,362	1,695	2,160	2,378	1,420	(Nov)	2,003
Humber Est.	2,875	1,425	1,795	2,480	1,405	(Dec)	1,996
Pulborough Levels	403	2,140	2,510	2,224	2,697	(Jan)	1,995
Swale Est.	2,353	2,040	1,846	1,311	2,057	(Dec)	1,921
Forth Est.	1,914	2,407	2,126	1,442	1,589	(Feb)	1,896
Dornoch Fth	2,307	1,761	1,831	1,406	1,682	(Oct)	1,797
Rutland Water	904	1,326	2,187	1,917	910	(Dec)	1,676
Lo. of Strathbeg	1,250	1,675	1,000	2,931	2,160	(Nov)	1,578
Southampton Water	1,388	1,239	1,677	1,675	1,372	(Dec)	1,470
Mere Sands Wood	600	1,500	1,369	2,664	985	(Nov)	1,424
Northern Ireland							
Lo. Neagh/Beg	2,155	1,576	2,915	1,805	1,669	(Feb)	2,024
Strangford Lo.	1,571	1,849	1,053	1,133	1,379	(Nov)	1,397

MALLARD
Anas platyrhynchos

International importance: 20,000**
Great Britain importance: 5,000
All-Ireland importance: 500

GB maximum: 170,954 Dec
NI maximum: 9,232 Sep

Trend	88-89	89-90	90-91	91-92	92-93
GB	87	81	79	75	75
NI	121	141	136	136	117

The peak count of Mallard recorded in Britain was remarkably similar to that in 1991-92, while the other monthly totals, except for the low January count, also closely matched those of the previous season. The population declined steadily over the last few seasons according to index values, with the 1992-93 value being the lowest of all 27 seasons used in the index. The Northern Ireland peak was the lowest in the last five winters, although the pattern of declining numbers as the winter progresses was maintained (Table 2).

That the December figure represented only around one third of the estimated GB population reflects the highly dispersed distribution of the species. No UK site approaches the level of international significance, and indeed no site in Great Britain reaches the GB 1% criterion. The qualifying level for including British sites in Table 26 has consequently been arbitrarily lowered to 2,000 birds. Owing to the number of hand-reared birds released for shooting each year, producing an accurate estimate of the Mallard population is difficult, and at least one release site where numbers are artificially high (Catchpenny Pool) appears in the table. No other site in Great Britain recorded in excess of 2,000 Mallard during 1992-93, whilst Ballysaggart Lough (500, September) was the only other site in Northern Ireland to hold 500 or more birds in 1992-93.

As part of the extensive research at Great Linford Gravel Pit complex, the effect of fish populations on Mallard duckling feeding success was examined (Phillips & Wright 1993). It had been recognised for some years that brood density was higher on natural lakes than man-made gravel pits, and the high numbers of fish present in these man-made waters was thought to be a major factor. The study confirmed that where fish numbers were high, the amount of invertebrate food available was low, and this had a negative effect on the ducklings' development. Conversely, where the fish population was relatively low, the feeding success of ducklings greatly improved, and as feeding bouts continued for longer periods and shorter distances were covered whilst feeding, it was suggested that this may have also reduced the risk of predation. Amongst other measures, the careful management of fish stocks may be important for the conservation of certain wildfowl.

Table 26. MALLARD: WINTER MAXIMA AT MAIN RESORTS

	88-89	89-90	90-91	91-92	92-93	(Mth)	Average
Great Britain							
Humber Est.	4,940	4,184	4,373	2,975	5,015	(Dec)	4,297
Ouse Washes	4,905	4,856	3,530	3,203	4,342	(Oct)	4,167
Wash	2,910	4,254	5,200	4,499	2,133	(Dec)	3,799
Morecambe Bay	4,670	4,496	3,400	3,504	2,780	(Nov)	3,770
Severn Est.	3,916	3,074	3,186	4,864	3,277	(Nov)	3,663
Martin Mere	3,900	3,600	4,170	3,000	2,835	(Jan)	3,501
Lower Derwent Ings	3,000	-	-	-	3,600	(Feb)	3,300
Dee Est.	4,105	3,505	1,947	3,720	2,681	(Oct)	3,192
Forth Est.	2,434	2,182	2,425	3,020	2,432	(Dec)	2,499
Solway Est.	2,666	2,185	1,511	2,422	2,183	(Oct)	2,193
Catchpenny Pool	-	-	-	1,397	2,751	(Sep)	2,074
Rutland Water	1,535	2,464	2,438	1,962	1,719	(Sep)	2,024
Northern Ireland							
Lo. Neagh/Beg	5,560	6,438	5,318	5,499	5,408	(Aug)	5,645
Lo. Foyle	2,000	1,889	2,309	1,799	1,596	(Sep)	1,919
Strangford Lo.	2,048	1,524	1,698	1,591	1,405	(Oct)	1,653
Upper Lo. Erne	722	1,115	694	-	448	(Jan)	745
Belfast Lo.	379	321	566	621	737	(Oct)	525

PINTAIL
Anas acuta

International importance:	700
Great Britain importance:	280
All-Ireland importance:	60

GB maximum: 20,993 Dec
NI maximum: 232 Dec

Trend	88-89	89-90	90-91	91-92	92-93
GB	189	188	163	194	149
NI	120	79	124	152	115

The peak count of Pintail in Britain in 1992-93 was low for the second time in three seasons. Although numbers returned to more normal levels for recent times in 1991-92, indices suggest that the wintering population is now smaller than during the 1990-91 low. The Northern Ireland peak was also considerably lower than 1991-92, though figures over the past five years have been variable. Many UK sites held far fewer Pintail in 1992-93 than expected, including all of the top six sites, with declines on the Wash and the Mersey being particularly dramatic (Table 27). Martin Mere, the Nene Washes, the Swale and Pagham Harbour were the only sites to support considerably more birds than their five year averages, although only the first site is of international importance for Pintail. In addition, Pulborough Levels (472, January) and the Blackwater (334, December) held in excess of 280 birds in 1992-93.

Since 1988-89, there appears to have been a regular peak of Pintail in October, albeit somewhat smaller than the regular December maximum (Table 3), first noted by Kirby *et al.* (1990). It is still uncertain why the numbers of birds present in November falls so sharply before the mid-winter peak, and where these birds go to. One possible explanation is an influx of Icelandic birds during October (P. Rose, pers comm), which then disperse within the UK or beyond, followed by a December or January peak in harsher weather conditions which includes birds from northern Europe. Unfortunately, insufficient ringing data are currently available to elucidate this phenomenon further.

Despite its apparently catholic nesting requirements in other countries, Pintail remain scarce breeders in the UK (Fox & Meek 1993). The wetlands of East Anglia are the main stronghold for the species, though many isolated records occur throughout Britain (Fox 1993b). To ensure the future of Pintail as a breeding species in the UK, further work is required to understand its ecological and nesting requirements, and to ensure that current sites are adequately protected and managed. Meek (1993) showed that over half the nest sites of Orkney birds had no formal protection and were potentially at risk from agricultural drainage. This indicates the need not only for site-based protection, but also for a conservation strategy that can deliver sympathetic wider-countryside measures.

Table 27. PINTAIL: WINTER MAXIMA AT MAIN RESORTS

	88-89	89-90	90-91	91-92	92-93	(Mth)	Average
International							
Dee Est.	8,435	11,945	8,706	10,001	7,605	(Oct)	9,338
Mersey Est.	4,288	8,000	3,200	6,089	3,504	(Dec)	5,016
Wash	6,541	2,757	1,910	3,509	317	(Sep)	3,007
Morecambe Bay	1,662	1,962	3,190	3,979	2,115	(Dec)	2,582
Ribble	830	1,621	556	6,507	1,850	(Nov)	2,273
Burry Inlet	1,800	2,306	1,784	1,657	1,137	(Dec)	1,737
Ouse Washes	1,228	1,818	1,332	1,969	1,745	(Feb)	1,618
Martin Mere	2,600	1,500	640	612	2,940	(Oct)	1,658
Solway Est.	1,575	1,092	2,208	1,200	1,124	(Oct)	1,440
Duddon Est.	2,200	873	830	1,189	1,260	(Feb)	1,270
North Norfolk Marshes	569	616	1,714	1,568	726	(Jan)	1,039
Medway Est.	927	700	1,243	1,233	473	(Dec)	915
Great Britain							
Nene Washes	520	883	128	55	970	(Jan)	511
Swale Est.	482	471	342	399	753	(Jan)	489
Pagham Hbr	245	173	839	174	778	(Nov)	442
Orwell Est.	291	506	-	528	432	(Dec)	439
Abberton Rsr	155	452	652	469	361	(Sep)	418
Cefni Est.	387	467	312	350	275	(Feb)	358
Stour Est.	278	182	320	362	380	(Feb)	304
Humber Est.	84	20	313	660	360	(Oct)	287
Northern Ireland							
Strangford Lo.	215	132	231	217	218	(Dec)	203
Lo. Foyle	125	42	90	134	35	(Feb)	85

GARGANEY
Anas querquedula

International importance:	20,000
Great Britain importance:	+*
All-Ireland importance:	+*

GB maximum:	25	Sep	Trend	not available
NI maximum:	0			

The customary peak count of Garganey was again made in autumn in 1992-93, although only numbering around half that of 1991-92. Although Garganey migrate to overwinter in sub Saharan Africa (del Hoyo *et al.* 1992), there were at least three records of overwintering birds in Britain, before numbers rose again in March. Nearly all records were from southeast England although birds seen at Morecambe Bay and on the Dee (England/Wales) were notable. Chew Valley Lake again held the highest number of birds (17, August, *cf.* 12 in 1991-92), whilst Rutland Water (7, September, *cf.* 6), the Dee (England/Wales) (6, September, *cf.* 0) and the Thames (6, September, *cf.* 0) all supported five or more birds. Spring and late summer counts at more sites would more accurately reflect the true situation.

SHOVELER
Anas clypeata

International importance:	400
Great Britain importance:	100
All-Ireland importance:	65

GB maximum:	7,873	Dec	
NI maximum:	153	Nov	

Trend	88-89	89-90	90-91	91-92	92-93
GB	171	186	221	213	160
NI	113	101	92	122	79

The peak British total of Shoveler was around 25% down on the peak 1991-92 figure, and was also the lowest count since the early 1980s. Annual indices confirm this picture, with the 1992-93 value being the lowest for 17 years. The Northern Ireland total was also particularly low, being the smallest total since regular counts began in 1985-86. The monthly variation in numbers during 1992-93 was considerably less than usual, with over 97% of the peak total present from September to December, declining to 75% by March (Table 3). The relatively mild weather conditions in November may have encouraged British birds to remain longer, delaying their usual autumn migration to the more southerly climes of France and Spain (Salmon 1986).

Shoveler exhibit a southeasterly bias in their distribution within the UK, favouring shallow, mesotrophic waters (Kirby & Mitchell 1993). A massive 46 sites qualify as nationally important for this species, the largest number for any species of wildfowl in the UK, although only five meet international criteria. Counts at many sites are quite variable from year to year, and numbers at any particular site rarely show any clear trends. However, numbers of Shoveler are currently much lower than two and three seasons ago, whilst numbers at Rutland also hint at a decline in the last five seasons. Counts on the Swale and at Dungeness were particularly high, with those at Stodmarsh, the Nene Washes, the Mid Avon Valley and, partly as a result of improved coverage, the Somerset Moors and Levels, also much higher than expected. Numbers at Fairburn Ings and on the Medway were only around half the normal number, while the very poor showing of only 68 birds at Woolston Eyes, especially in view of the 420 recorded there in 1991-92, results from the drainage last season of what was previously a regionally and nationally important site (Kirby & Owen 1993). A further 10 sites held 100 or more Shoveler in 1992-93, namely the Cotswold Water Park East (214, September), Arlington Reservoir (176, December), Hule Moss (174, October), Hardley Flood (169, September), Poole Harbour (164, January), the Blackwater (119, February), Kempton Reservoir East (101, January), Ranworth and Cockshoot Broads (101, October), the North West Solent (100, February) and the Lower Derwent Ings (100, December).

Table 28. SHOVELER: WINTER MAXIMA AT MAIN RESORTS

	88-89	89-90	90-91	91-92	92-93	(Mth)	Average
International							
Abberton Rsr	418	829	1,085	608	520	(Sep)	692
Ouse Washes	523	696	625	567	724	(Feb)	627
Rutland Water	729	372	680	490	362	(Nov)	527
Chew Valley Lake	475	490	465	630	435	(Nov)	499
Lo. Leven	540	285	540	576	448	(Oct)	478
Great Britain							
Swale Est.	348	224	276	447	591	(Mar)	377
Wraysbury Rsr	-	601	233	194	237	(Oct)	316
Dungeness RSPB	186	-	-	191	435	(Nov)	271
Woolston Eyes	167	300	260	420	68	(Oct)	243
Fiddlers Ferry Lagoons	220	464	148	200	110	(Sep)	228
Burry Inlet	348	149	172	219	226	(Dec)	223
Thames Est.	258	300	158	134	144	(Feb)	200
Fairburn Ings	115	282	203	145	90	(Oct)	167
Leighton Moss	155	166	140	205	148	(Sep)	163
Knight/Bessborough Rsrs	-	-	-	-	160	(Nov)	160
Stodmarsh NNR	-	100	102	132	300	(Dec)	159
Medway Est.	166	159	250	143	75	(Feb)	159
Nene Washes	81	247	97	41	318	(Mar)	157
Grafham Water	63	146	263	207	108	(Jan)	157
Thrapston GP	252	134	41	117	222	(Sep)	153
King George VI Rsr	333	87	26	117	201	(Mar)	153
Pulborough Levels	124	156	93	118	261	(Jan)	150
Blithfield Rsr	28	89	40	441	146	(Oct)	149
Attenborough GP	124	126	286	95	112	(Dec)	149
North Warren/Thorpeness Mere	-	-	-	98	184	(Feb)	141
Walthamstow Rsr	77	112	214	107	160	(Sep)	134
Fleet/Wey	37	148	149	208	121	(Dec)	133
Avon Valley (Mid)	46	74	121	157	245	(Jan)	129
Blagdon Lake	92	136	208	64	144	(Nov)	129
Cheshunt GP	110	180	115	150	90	(Feb)	129
Stockers Lake	-	-	-	100	156	(Dec)	128
Crouch/Roach Est.	124	126	102	155	134	(Jan)	128
Kingsbury/Coton Pools	90	154	92	127	172	(Nov)	127
Hanningfield Rsr	30	101	112	216	151	(Oct)	122
Somerset Levels	3	154	63	88	291	(Feb)	120
Langstone Hbr	212	88	87	78	133	(Dec)	120
Willen Balance Lake	143	87	154	73	139	(Dec)	119
North Norfolk Marshes	63	119	93	139	173	(Sep)	117
Pitsford Rsr	87	107	112	158	122	(Sep)	117
Swithland Rsr	56	45	72	181	179	(Sep)	107
Swillington Ings	80	84	-	173	92	(Dec)	107
Tees Est.	209	78	59	98	86	(Sep)	106
Solway Est.	117	191	83	113	27	(Nov)	106
Barn Elms Rsr	100	107	110	116	98	(Feb)	106
Tophill Low Rsrs	117	92	96	51	148	(Sep)	101
Ashford Common Waterworks	-	-	-	-	100	(Feb)	100
Northern Ireland							
Lo. Neagh/Beg	86	196	169	126	56	(Mar)	127
Strangford Lo.	140	109	95	143	130	(Dec)	123

RED-CRESTED POCHARD
Netta rufina

International importance:		200
Great Britain importance:		introduced
All-Ireland importance:		introduced

GB maximum: 134 Nov
NI maximum: 0

Trend   not available

Numbers of this exotic looking duck from October onwards were around double those recorded in 1991-92. Although some continental birds may visit Britain during winter, numbers are likely to be small. The majority of British birds are resident, with small populations having arisen from escaped birds (Cox 1986). The large increase in 1992-93, in excess of known increases at some sites (e.g. Baatsen 1990), is thus likely to be a result of more observers having been prepared to record Red-crested Pochards. Although birds are known to move quite regularly between adjacent sites (Baatsen 1990), giving rise to the possibility of double-counting, there are also likely to be birds that go undetected by the WeBS scheme. Previous estimates have suggested that the British population numbered less than 100 birds (Cox 1986, Delany 1993c), but given known increases at some of the key sites, this figure may have risen as high as 150 at present and seems set to continue. Most birds were located at or around known strongholds in the south and south east, although there were records from Clwyd and several sites in Scotland, including the Orkneys. The Cotswold Water Park held around half the total, with 52 in the West (*cf.* 31 in 1991-92) and 13 in the East (*cf.* 9), both in November. Pensthorpe Lakes (48, November, *cf.* 33), Paultons Bird Park (42, January, *cf.* 0) and the Serpentine (11, March, not counted in 1991-92) were the other sites to hold more than 10 birds in 1992-93.

POCHARD
Aythya ferina

International importance: 3,500
Great Britain importance: 440
All-Ireland importance: 400

GB maximum: 36,490 Jan
NI maximum: 23,928 Jan

Trend	88-89	89-90	90-91	91-92	92-93
GB	109	104	99	95	91
NI	141	128	131	139	102

Peak numbers in Northern Ireland declined steeply in 1992-93, following a series of consistently high midwinter counts. Index values for recent seasons confirm this picture. The large decline at Loughs Neagh and Beg, which supports around 95% of the total, resulted in the lowest count at the site since 1986-87. Numbers here had increased markedly from the mid 1980s to represent well in excess of 10% of the international population in recent seasons. This is somewhat anomalous given the general stability of both the British and the northwest European populations in the last 15 years (Kirby *et al.* in press, a, Monval & Pirot 1989). Given the magnitude of the decline in 1992-93, it seems likely that several factors contributed to the population level of Pochard and other waterfowl at the site. There is some evidence to suggest that there is an inverse relationship between fish numbers and numbers of *Aythya* species (Winfield *et al.* 1992) and it is possible that such a relationship may hold for the low numbers of Pochard at Loughs Neagh and Beg in 1992-93. Also, numbers of Pochard at the site fluctuate markedly throughout the winter (Table 4), with the maximum count often being considerably larger than the other monthly counts. Thus, another possible explanation is that there were fewer immigrants during the mild weather in 1992-93 or that the peak was particularly short-lived and simply did not coincide with WeBS count dates. It will be necessary to keep a close eye on numbers at this site in future seasons.

Peak numbers of Pochard in Britain, which has generally held slightly fewer birds than Northern Ireland in recent years, were about average in 1992-93 compared with recent winters. Index values, however, point to a steady decline of over 15% over the last five years, again, contrary to the international picture (Monval & Pirot 1989).

Numbers of Pochard at many of the key sites are rather variable (Table 29). Most of the top 10 sites recorded less than average numbers, with declines at Abberton Reservoir, the Loch of Harray, Rostherne Mere and at the Cotswold Water Park as a whole (Delany 1993b) being particularly dramatic. However, numbers on the Ouse Washes and on the Fleet/Wey have increased markedly, with those at the latter site having done so for the fifth season in succession. Numbers at Eyebrook Reservoir, although having fluctuated in recent seasons, were exceptional in 1992-93, representing the fifth largest count at any site in the UK. The Nene Washes (1,014, November) and the Thames Estuary (432, October) were the only other sites to support more than 430 Pochard in 1992-93.

Table 29. POCHARD: WINTER MAXIMA AT MAIN RESORTS

	88-89	89-90	90-91	91-92	92-93	(Mth)	Average
International							
Lo. Neagh/Beg	39,811	36,380	40,928	38,998	23,367	(Jan)	35,897
Great Britain							
Abberton Rsr	2,739	2,271	4,064	2,058	1,420	(Nov)	2,510
Ouse Washes	1,129	2,964	1,135	1,596	3,279	(Feb)	2,021
Severn Est.	2,026	1,742	1,616	1,666	1,623	(Feb)	1,735
Lo. of Boardhouse	723	1,327	1,594	1,864	1,135	(Dec)	1,329
Kingsbury/Coton Pools	1,408	1,387	1,099	1,275	1,085	(Nov)	1,251
Lo. of Harray	1,372	1,011	2,245	846	452	(Oct)	1,185
Cotswold WP East	1,352	1,147	1,113	1,462	691	(Dec)	1,153
Cotswold WP West	1,538	1,329	1,046	811	1,015	(Nov)	1,148
Rostherne Mere	1,151	120	2,703	635	466	(Feb)	1,015
Loch Leven	770	1,510	895	701	900	(Dec)	955
Poole Harbour	+685	+1,020	1,311	1,026	438	(Dec)	896
Chew Valley Lake	2,450	625	260	370	580	(Oct)	857
Fleet/Wey	416	683	749	883	1,051	(Jan)	756
Rutland Water	532	626	1,218	533	656	(Jan)	713
Avon Valley (Mid)	655	469	657	720	836	(Jan)	667
Stanton Harcourt GP	340	470	1,243	237	645	(Jan)	587
Kilconquhar Lo.	384	866	544	752	253	(Jan)	560
Eyebrook Rsr	58	574	451	373	1,313	(Sep)	554
Dungeness RSPB	247	-	-	955	405	(Sep)	536
Humber Est.	545	39	700	429	900	(Jan)	523
Lo. Ore	659	-	-	223	658	(Sep)	513
Hanningfield Rsr	414	316	468	667	411	(Aug)	455
Arundel WWT	372	462	582	458	391	(Feb)	453

+ collated from county records by S. Aspinall.

TUFTED DUCK
Aythya fuligula

International importance:	7,500
Great Britain importance:	600
All-Ireland importance:	400

GB maximum:	52,639	Dec	
NI maximum:	19,006	Jan	

Trend	88-89	89-90	90-91	91-92	92-93
GB	126	99	109	99	102
NI	105	111	105	117	109

Numbers of Tufted Duck in Britain have remained relatively stable since the early 1970s, following an earlier increase (Table 5), and although the peak count was some 2,000 greater than that of 1991-92 and slightly higher than in recent seasons, there is as yet no firm indication of a genuine increase. The peak total for Northern Ireland, however, was some 27% down on the previous season. Like Pochard, peak numbers in Northern Ireland are subject to large fluctuations, being particularly influenced by numbers at Loughs Neagh and Beg, and though this appears to be a sharp reduction it still remains well within the bounds of previous fluctuations. Winfield *et al.* (1992) suggested that the populations of Great Crested Grebe and Tufted Duck at Lough Neagh were inversely linked due to the effect of variations in fish stocks, which provide prey items for grebes and alter benthic structure, affecting diving ducks. The numbers of these species in 1992-93 lend support to this theory. Seasonal variations in Great Britain remained relatively small, with around 80% of the peak total present until March, whilst in Northern Ireland an early influx of birds occurred in October, perhaps related to the particularly cold temperatures experienced in Scotland at that time which may have caused the breeding population to vacate this area sooner than usual.

As for Pochard, numbers of Tufted Duck were below average at many of the key sites (Table 30), with steady declines recorded at Abberton Reservoir and Rutland Water in recent winters. Conversely, numbers at both Hanningfield and Pitsford Reservoirs have grown steadily, with the count of moulting birds at the former site representing the third highest at any site in the UK in 1992-93. The Ouse Washes (1,347, February), Windermere (948, November) and South Muskham and North Newark Gravel Pits (689, February) also recorded numbers in excess of 600 Tufted Duck in 1992-93.

Suter & Van Eerden (1992) examined two incidents of mass starvation of wintering ducks in 1986, involving over 700 Tufted Duck on a section of the River Rhine in Switzerland and over 1,000 in the Dutch Wadden Sea. A prolonged period of cold weather occurred across Europe in February, and a limited supply of Zebra Mussels *Dressiana polymorpha* and Blue Mussels *Mytilus edulis* at the end of the winter period 700

Tufted Duck on a section of the River Rhine in Switzerland and over 1,000 in the Dutch Wadden Sea. A prolonged period of cold weather occured across Europe in February and a limited supply of Zebra Mussels *Dressiana polymorphus* and Blue Mussels *Mytilus edulis* at the end of the winter period appears to have been the primary cause of mortality; with spring migration so close, the birds seemed reluctant to leave the area to seek alternative food supplies further from their breeding grounds, preferring instead to "sit out" the harsh weather. In most circumstances this would have been the most energetically efficient strategy, though the prolonged period of the harsh conditions on this occasion caused large scale mortality. Tufted Duck casualties were less than those suffered by Pochard, probably due to their superior diving ability.

Table 30. TUFTED DUCK: WINTER MAXIMA AT MAIN RESORTS

	88-89	89-90	90-91	91-92	92-93	(Mth)	Average
International							
Lo. Neagh/Beg	16,642	29,393	22,278	25,283	18,078	(Dec)	22,335
Great Britain							
Abberton Rsr	3,987	4,387	3,550	2,428	1,724	(Dec)	3,215
Lo. Leven	3,180	2,700	3,120	4,064	2,500	(Sep)	3,113
Rutland Water	5,582	3,709	2,097	2,397	1,723	(Sep)	3,102
Lo. of Harray	1,920	1,992	1,643	1,570	1,920	(Nov)	1,809
Kingsbury/Coton Pools	1,405	1,794	1,431	1,177	1,471	(Dec)	1,456
Hanningfield Rsr	655	530	537	1,463	2,055	(Aug)	1,048
Wraysbury GP	1,447	683	470	1,229	1,291	(Oct)	1,024
Kilconquhar Lo.	1,570	1,250	725	596	702	(Jan)	969
Walthamstow Rsr	721	721	1,589	1,035	663	(Jan)	946
King George V Rsr	430	530	2,500	862	380	(Sep)	940
Besthorpe/Girton GP	1,100	557	1,801	678	509	(Jan)	929
Severn Est.	990	997	817	786	967	(Dec)	911
Pitsford Rsr	501	697	764	766	1,107	(Sep)	767
Cotswold WP West	1,322	464	593	694	627	(Mar)	740
Avon Valley (Mid)	705	472	657	746	739	(Jan)	664
Inner Moray Firth	208	640	777	595	800	(Jan)	604
Heaton Park Rsr	370	76	426	1,312	820	(Jan)	601
Northern Ireland							
Upper Lo. Erne	637	620	349	-	588	(Feb)	549

SCAUP
Aythya marila

International importance: 3,100
Great Britain importance: 110
All-Ireland importance: 30*

GB maximum:	3,734	Feb	Trend	not available
NI maximum:	3,955	Mar		

The British peak in 1992-93 remained much the same as the previous season, though this probably represents an under count since numbers recorded at key sites, such as the Solway, are heavily influenced by local counting conditions. The peak count in Northern Ireland rose for the fourth consecutive year, exceeding the total for the rest of the UK. Monthly fluctuations showed a steady build up throughout the winter, with a pronounced March peak, perhaps representing staging birds returning to Iceland (Table 4).

The Solway Estuary has held the largest numbers of birds of any site in Britain since the steady decline of the Forth Estuary Scaup population, which numbered 25,000 birds in the late 1960s and early 1970s (Owen *et al.* 1986). It is the only site in the UK currently of international importance for this species. Numbers at Loughs Neagh and Beg remained high following the recent increase. The Moray Firth population, which includes three nationally important sites (the Cromarty, Dornoch and Inner Moray Firths) has remained relatively stable for the last 15 years (Evans 1993). Numbers at Carlingford Lough and Belfast Lough continue to rise in line with the Northern Ireland population in general, although the numbers of birds at the latter site remains comparatively small. Numbers at English resorts tend to fluctuate widely, with the count on the Wash being notable. The Clwyd Estuary (108, January), Rough Firth (74, February) and Wigtown Bay (60, January) also held over 40 birds in 1992-93.

Table 31. SCAUP: WINTER MAXIMA AT MAIN RESORTS

	88-89	89-90	90-91	91-92	92-93	(Mth)	Average
International							
Solway Est.	3,092	1,562	3,803	++5,400	1,686	(Feb)	3,109
Great Britain							
Lo. Indaal	1,230	442	660	1,430	1,120	(Jan)	976
Forth Est.	762	135	381	185	188	(Dec)	330
Loch Ryan	200	409	200	300	500	(Feb)	322
Cromarty Fth	151	+126	+247	+211	293	(Feb)	206
Dornoch Fth	266	149	+368	140	81	(Jan)	201
Lo. of Harray	219	164	240	95	132	(Oct)	170
Inner Clyde	174	267	144	146	108	(Jan)	168
Lo. of Stenness	47	19	194	243	233	(Jan)	147
Inner Moray Fth	149	84	195	228	41	(Feb)	139
Irvine to Saltcoats	-	-	-	-	113	(Feb)	113
Irvine/Garnock Est.	45	18	73	200	55	(Jan)	78
Eden Est.	79	68	32	0	100	(Jan)	56
Wash	31	3	35	18	141	(Nov)	46
Humber Est.	65	0	128	5	30	(Jan)	46
Northern Ireland							
Lo. Neagh/Beg	2,150	1,215	1,539	3,516	3,384	(Mar)	2,361
Carlingford Lo.	140	150	352	500	570	(Feb)	342
Belfast Lo.	19	29	43	74	121	(Feb)	57

+ RSPB/BP studies
++ WWT/WAS studies

EIDER
Somateria mollissima

International importance: 20,000
Great Britain importance: 750
All-Ireland importance: 20*

| GB maximum: | 27,567 | Sep |
| NI maximum: | 663 | Oct |

Trend: not available

The peak count of Eider in Britain in 1992-93 was considerably less than totals recorded in the previous four seasons. It is, however, strongly influenced by the numbers recorded on the Tay Estuary, which are themselves variable and, as with many counts of sea-duck, subject to the prevailing conditions. The count of 250 at that site in 1992-93 was very much smaller than the 20,000-30,000 of recent seasons. The peak of 27,567 birds represented only one third of the British population, although, if numbers on the Tay are taken into account, well over half the population is found on WeBS sites. This is a comparatively large proportion of the true figure compared with many other sea-duck, reflecting the more inshore distribution of Eiders and their use of more accessible sites, notably east coast Scottish estuaries. The Shetlands, Orkneys and Western Isles support the bulk of the uncounted birds (Kirby et al. 1993). The considerable variability in numbers that arises directly as a result of censusing difficulties precludes the meaningful use of WeBS data for analyses of monthly fluctuations or long-term trends, although recent work suggests the British population is stable or increasing (Kirby et al. 1993).

The predominance of northeast English and east Scottish sites in Table 32 reflects the wintering distribution in the UK, with the Clyde and Morecambe Bay representing the only significant concentrations outside this area. Numbers on the Firth of Forth and Morecambe Bay were higher than average while those on the Irvine and Garnock Estuary appear to have risen dramatically in recent years. Other British sites which held more than 750 Eider in 1992-93 were Troon shore (911, September), Turnberry to Dipple Shore (900, February), the Wash (873, February), Burghead to Hopeman (850, October) and Greater Cumbrae (831, February), whilst the Bann Estuary (20, October) was the only other site in Northern Ireland to support 20 or more birds.

Bustnes (1993) described how single female Eider exploit vigilance of others during the brood-rearing period. Single birds would often feed close to creches of ducklings being supervised by another female. The need for extreme vigilance by the brood-caring female is used by other females to spend a greater proportion of their time feeding, and significantly less time being vigilant, enabling them to be forewarned of any potential attacks by predators, a behaviour that has evolved since no brood territories are maintained.

Table 32. EIDER: WINTER MAXIMA AT MAIN RESORTS

	88-89	89-90	90-91	91-92	92-93	(Mth)	Average
International							
Tay Est.	-	30,000	20,300	(5,000)	(250)	(Sep)	25,150
Great Britain							
Forth Est.	3,977	10,798	7,836	6,219	9,375	(Mar)	7,641
Morecambe Bay	5,773	7,604	8,183	8,089	7,509	(Nov)	7,432
Inner Clyde	4,384	4,674	2,939	4,164	2,584	(Oct)	3,749
Irvine to Saltcoats	-	-	-	-	2,500	(Jan)	2,500
Lindisfarne	2,300	2,000	2,600	1,750	2,380	(Jul)	2,206
Montrose Basin	2,000	2,960	2,100	1,898	1,316	(Mar)	2,055
Ythan Est.	1,315	1,013	2,322	2,586	2,065	(Jul)	1,860
Seahouses to Budle Pt	-	310	-	3,200	1,642	(Sep)	1,717
Don Mouth to Blackdog	1,000	2,000	400	2,500	2,000	(Sep)	1,580
Blyth to Newbiggin	-	-	-	-	900	(Jan)	900
Irvine/Garnock Est.	369	67	800	1,600	1,500	(Jan)	867
Northern Ireland							
Belfast Lo.	137	440	1,227	1,107	641	(Oct)	710
Outer Ards	410	377	303	299	166	(Jan)	311
Larne Lo.	42	75	37	88	59	(Sep)	60
Lo. Foyle	2	21	3	4	106	(Sep)	27

LONG-TAILED DUCK
Clangula hyemalis

International importance: 20,000
Great Britain importance: 230
All-Ireland importance: +*

GB maximum:	1,198	Jan	Trend	not available
NI maximum:	35	Dec		

WeBS counts detect only a small proportion of the total Long-tailed Duck population wintering around the British Isles, since, like many other species of sea-duck, it is difficult to count unless viewing conditions are at least good. Specific surveys, such as the RSPB/BP monitoring of the Moray Firth, record much larger numbers and this area remains the major stronghold for this species in the UK, holding around 39% of the estimated UK winter population (Evans 1993). Even these counts must be regarded as minimum values, and a "best estimate" (the sum of the peak monthly totals recorded at each site) for 1992-93 in the Moray Firth has been calculated at 12,170 birds. The favoured area is the coast from Burghead to Culbin, including an important roost site in Burghead Bay.

Since counts at many sites probably under-record their significance for Long-tailed Duck, Table 33 has been extended to include all sites with five year average maxima of 70 or more birds, although it should be noted that only those regularly holding over 230 birds qualify as nationally important. Away from the Moray Firth, the Forth is the only location to have held over 1,000 birds in the last five years, and although numbers have been well down on this peak in recent seasons, it remains well above its nearest rival. Lindisfarne, the only wintering site of national importance in England, held low numbers in 1992-93. Other sites holding 70 or more birds were Loch Branahuie (80, October) and North Norfolk Marshes (70, March), the most southerly site regularly supporting notable numbers of birds. As usual, only small numbers were recorded in Northern Ireland.

Table 33. LONG-TAILED DUCK: WINTER MAXIMA AT MAIN RESORTS

	88-89	89-90	90-91	91-92	92-93	(Mth)	Average
Great Britain							
Moray Fth	10,500	+6,270	+8,037	+9,300	+11,246	(Feb)	9,071
Forth Est.	1,037	465	451	640	491	(Mar)	617
Lindisfarne	800	294	420	(42)	83	(Feb)	399
Water Sound	++219	206	365	-	240	(Mar)	257
Additional sites							
Sound of Taransay	134	(0)	-	-	112	(Nov)	123
Lo. Fleet	189	0	26	212	123	(Jan)	110
Cromarty Fth	7	0	147	269	80	(Feb)	101
Broad Bay	-	100	-	-	-		100
Lo. of Stenness	141	77	75	53	114	(Jan)	92
Lo. of Harray	133	29	111	53	57	(Dec)	77

+ *RSPB/BP studies*
++ *from Christer (1989)*

COMMON SCOTER
Melanitta nigra

International importance:	8,000
Great Britain importance:	230
All-Ireland importance:	40*

GB maximum: 2,293 Mar
NI maximum: 499 Dec

Trend   not available

Although separate accounts for Common and Velvet Scoter are provided here, it is important to remember that, even when counting flocks under the best of viewing conditions, the species are not always distinguishable and consequently the figures given in these accounts should be interpreted with caution.

Unusually low numbers of Common Scoter were recorded by WeBS in 1992-93, both in Britain and Northern Ireland. Counts are often hampered by the prevailing weather, and the fact that some birds reside up to 7 km offshore (Green & Elliott 1993) further complicates censusing. It has been estimated that shore counts may detect less than a fifth of the birds present (Owen *et al.* 1986). Nevertheless, the fall in numbers in 1992-93 is particularly large, although it is difficult to say with any confidence whether this represents a genuine decrease.

The continuing survey of wintering birds in Cardigan Bay again detected high numbers in 1992-93, the maximum count achieved by aerial surveys (Green & Elliott 1993). Although numbers have fluctuated quite dramatically in this area, figures suggest it holds at least 20% of the British population. The Common Scoter that winter in the Irish Sea tend to be highly mobile, with large scale changes in wintering areas (A. Webb, *in litt.*), and it is possible that birds moved to a previously unrecognized site. However, the low counts at many sites, notably the three most important in the UK, two of which are counted using specific methodology aimed at recording sea-ducks, lend weight to the possibility of a genuine decline in 1992-93.

Table 34 lists the key sites for Common Scoter in the UK, although the difficulties of counting sea-ducks means that the significance of many sites for these species especially will be under-recorded by WeBS. The count on the Forth was one of the few to exceed its average number, whilst Ythan to Collieston also recorded comparatively large numbers. The major site in Northern Ireland, Dundrum Bay, suffered a large drop in numbers, with only 25% of the average peak being recorded. Ythan to Collieston (454, September) and Sound of Tarransay (250, January) were the only other sites to support more than 230 Common Scoter in 1992-93.

Table 34. COMMON SCOTER: WINTER MAXIMA AT MAIN RESORTS

	88-89	89-90	90-91	91-92	92-93	(Mth)	Average
International							
North Cardigan Bay	(1,735)	(4,100)	++1,569	++10,397	++5,650	(Feb)	5,872
Moray Firth	-	-	+1,818	+9,933	+2,197	(Mar)	4,649
Dundrum Bay	2,580	1,150	2,480	2,962	498	(Dec)	1,934
Forth Est.	1,958	1,047	1,735	1,330	1,773	(Feb)	1,569
Great Britain							
Eden Est.	400	2,100	700	(0)	1,000	(Jan)	1,050
North Norfolk Marshes	1	31	531	3,207	764	(Dec)	907
Lindisfarne	1,000	500	1,100	600	436	(Jan)	727
Tay Est.	500	400	2,150	0	0		610
Carmarthen Bay	-	2,050	3	2	26	(Oct)	520
Wash	510	225	1,034	123	48	(Dec)	388
Thames Est.	415	170	87	392	250	(Mar)	263
Red Wharf Bay	112	57	922	70	100	(Mar)	252
Cresswell/Chev'ton Burn	300	(2)	-	400	20	(Oct)	240

+ RSPB/BP studies
++ Elliott & Green (1993)

VELVET SCOTER
Melanitta fusca

International importance: 2,500
Great Britain importance: 30*
All-Ireland importance: +*

GB maximum: 294 Mar
NI maximum: 0

Trend: not available

As for Common Scoter, the difficulties involved in monitoring sea-ducks and especially the problem of species separation, mean that figures given here must be used with caution. However, given that these birds are by far the least common of the two, they are undoubtedly overlooked in large, mixed flocks, and true numbers are almost undoubtedly higher.

The peak count of Velvet Scoter during 1992-93 was lower than 1991-92, as to be expected given the large drop in Common Scoter, with only around 10% of the British population being recorded. No birds were specifically identified in Northern Ireland. Few sites hold Velvet Scoter with any regularity and, even here, numbers often fluctuate widely. No UK sites are of international importance for this species, although the Eden Estuary and the Moray Firth have both held large numbers at least once in recent seasons. The winter survey of the Moray Firth recorded over 1,000 Velvet Scoter for the first time in eight years, though this is still well below numbers recorded in the early 1980s (Evans 1993). Table 35 lists all sites with a five year average of 30 or more Velvet Scoter, with the Wash (54, December) being the only other site to hold in excess of this figure in 1992-93.

Table 35. VELVET SCOTER: WINTER MAXIMA AT MAIN RESORTS

	88-89	89-90	90-91	91-92	92-93	(Mth)	Average
Great Britain							
Eden	500	2,400	400	(0)	4	(Jul)	826
Moray Firth	-	-	+328	+487	+1,039	(Mar)	618
Forth Est.	47	118	84	180	290	(Mar)	144
North Cardigan Bay	-	-	40	26	38	(Feb)	35
North Norfolk Marshes	0	0	3	167	4	(Nov)	35

+ RSPB/BP studies

GOLDENEYE
Bucephala clangula

				International importance:				3,000
				Great Britain importance:				170
				All-Ireland importance:				110

			Trend	88-89	89-90	90-91	91-92	92-93
GB maximum:	16,388	Feb	GB	102	96	130	122	119
NI maximum:	14,729	Jan	NI	83	77	99	107	94

The peak count of Goldeneye, although slightly lower than the 1991-92 maximum, remained at the high levels recorded by WeBS in 1990-91; around 25-30% higher than counts in previous seasons. This increase is also reflected in the annual indices, with numbers having remained relatively stable from the late 1970s to the mid 1980s, before a gradual increase (Table 5). The picture in Northern Ireland is very similar, with numbers having increased during the late 1980s, although the 1992-93 peak was rather less than the 16,000 recorded in the previous season. Goldeneye numbers generally increase steadily through the winter to reach mid-winter peaks in February in Britain and in December, January or February in Northern Ireland (Tables 3 & 4). 1992-93 numbers mirrored this pattern, although there was a large and unexplained fall in December in Northern Ireland prior to the January peak.

Numbers at many of the key sites given in Table 36 are characterised by a peculiarly large count in one or two seasons; several sites with average maxima of 200-300 birds having held in excess of 500 or 600 birds at least once in the last five winters. Peak numbers at Loughs Neagh and Beg have remained remarkably constant in recent years and represent the only internationally important concentration of Goldeneye in the UK, holding over 90% of the Northern Ireland total. Numbers in the Forth Estuary remained relatively high, and appear to have recovered after a large drop in numbers during the late 1970s (Owen *et al.* 1986), whilst Maidens Harbour, Strangford Lough and the Doon Estuary also held comparatively large numbers. Many of the remaining sites, however, held slightly fewer than might be expected from their five year averages. Other sites which supported numbers in excess of 170 Goldeneye in 1992-93 were the Humber Estuary (301, December), Poole Harbour (213, January) and Loch of Harray (194, February).

Duncan & Marquiss (1993) showed that counts at roost sites in northeast Scotland remained relatively constant throughout the winter, though observations at feeding sites suggested that, as the winter progressed, birds moved away from lochs to feed on the lower reaches of rivers where food supplies were less depleted. It was also argued that adult males have a competitive advantage over females where food is concentrated in deep or fast-flowing water due to their ability to dive for longer periods and hence make more use of these sites. Females and young birds only utilised these habitats when food elsewhere was in short supply.

Table 36. GOLDENEYE: WINTER MAXIMA AT MAIN RESORTS

	88-89	89-90	90-91	91-92	92-93	(Mth)	Average
International							
Lo. Neagh/Beg	12,239	11,408	13,591	13,565	13,748	(Jan)	12,910
Great Britain							
Forth Est.	1,608	991	1,831	2,451	2,167	(Jan)	1,810
Inner Moray Fth	682	+680	+993	+718	+820	(Feb)	779
Abberton Rsr	1,002	362	707	673	547	(Mar)	658
Maidens Hbr/Turnberry	-	625	462	312	900	(Feb)	575
Inner Clyde	607	413	609	541	575	(Mar)	549
Morecambe Bay	430	480	425	538	384	(Mar)	451
Tweed Est.	408	360	351	590	329	(Feb)	408
Rutland Water	385	345	505	412	369	(Mar)	403
Windermere	256	292	329	501	357	(Jan)	347
Blackwater Est.	172	518	424	236	330	(Jan)	336
Lo. of Skene	-	415	358	144	404	(Mar)	330
Loch Leven	330	310	265	330	300	(Feb)	307
Tay Est.	41	214	210	677	274	(Jan)	283
R. Tweed: Coldstream	315	238	228	295	334	(Mar)	282
Doon Est.	142	200	324	236	500	(Jan)	280
Solway Est.	127	213	504	174	294	(Feb)	262
Irvine to Saltcoats	-	-	-	-	250	(Jan)	250
Colne Est.	181	150	477	227	202	(Jan)	247
Dipple	-	-	462	92	140	(Feb)	231
Lo. of Stenness	189	202	259	225	242	(Jan)	223
Ayr Hbr to Greenan Castle	-	-	-	237	176	(Mar)	207
Kilconquhar Lo.	77	644	31	46	177	(Jan)	195
Lo. of Strathbeg	68	249	170	338	67	(Dec)	178
Ayr to Prestwick	157	226	215	176	113	(Nov)	177
Cromarty Fth	133	209	158	148	212	(Feb)	172
Northern Ireland							
Belfast Lo.	320	578	634	283	375	(Feb)	438
Strangford Lo.	289	240	290	373	531	(Nov)	345
Larne Lo.	122	113	126	316	234	(Mar)	182
Upper Lo. Erne	182	177	206	-	156	(Nov)	180

+ RSPB/BP studies

SMEW
Mergus albellus

International importance: 150
Great Britain importance: 2*
All-Ireland importance: +*

GB maximum: 108 Feb
NI maximum: 1 Nov/Feb

Trend: not available

Numbers of Smew in 1992-93 were about average for a mild winter, with few birds until mid-winter and a late winter peak. As usual, birds were concentrated in southeast England, although there were several multiple records from Scotland and also Wales. Wraysbury Gravel Pits, the principal site in recent years, again claimed the honours, with a notable 39 birds in February (*cf.* 17 in 1991-92) while the count of 14 at Llyn Bran Bylchan in March (not counted in 1991-92) was similarly noteworthy in view of it being the largest single species count at the site during 1992-93. Dungeness RSPB Reserve (11, February, *cf.* 9), Eyebrook Reservoir (6, February, *cf.* 2), Pagham Harbour (5, January, *cf.* 0), Staines Moor Gravel Pits (5, January, not counted in 1991-92) and Wraysbury Reservoir (5, February, *cf.* 0) were the other sites to support five or more birds in 1992-93.

RED-BREASTED MERGANSER
Mergus serrator

International importance:	1,000
Great Britain importance:	100
All-Ireland importance:	20*

GB maximum:	4,033	Feb
NI maximum:	825	Nov

Trend	88-89	89-90	90-91	91-92	92-93
GB	165	104	131	123	129
NI	115	109	105	85	110

Large numbers of Red-breasted Merganser were recorded in Britain in 1992-93, with the peak British count exceeding 4,000 for the first time since the mid 1980s, although this still represents less than half the total estimated population (Kirby in prep.). The Northern Ireland total was also considerably higher than in the two previous seasons. Annual indices show fluctuating numbers in recent years, having declined somewhat from peak figures in the 1980s (Table 5).

A study of the Red-breasted Merganser population on the river North Esk revealed that, despite April shooting, the number of breeding pairs remained relatively stable. This suggests that the congregations on coastal waters in April prior to dispersal included a considerable number of migrant birds which replaced any resident birds which were shot (Marquiss & Duncan 1993). The study also suggested that although the breeding density of birds was not correlated with the density of their main spring food source, Salmon parr *Salmo salar*, it may have been correlated with food availability, since parr are easier to catch in slower moving water without a rocky substrate. It is suggested that the timing of breeding coincides with an abundance of juvenile fish and large aquatic invertebrates which form the main diet of young ducklings.

Following the large national total, many individual sites also registered large numbers of Red-breasted Merganser in 1992-93, with five of the top six sites holding well above average numbers. The high counts on both the Firth of Forth and the Duddon were notable in view of the static peaks in recent winters. Numbers on the Tay were low for the second winter in succession and although numbers were also low on the Wash, they appear to fluctuate quite widely at this site. More worrying is the continued demise of birds at Lindisfarne which will undoubtedly soon cease to be of national importance for this species. Other British sites supporting more than 100 Red-breasted Merganser in 1992-93 were Arran (132, October), Pagham Harbour (118, March) and the Solway (116, December). In Northern Ireland, Loughs Neagh and Beg (33, August) and the Bann Estuary (31, December) held more than 30 birds in 1992-93.

Table 37. RED-BREASTED MERGANSER: WINTER MAXIMA AT MAIN RESORTS

	88-89	89-90	90-91	91-92	92-93	(Mth)	Average
International							
Inner Moray Fth	+1,777	+1,255	+1,440	+1,789	+1,315	(Feb)	1,515
Great Britain							
Forth Est.	437	472	478	459	726	(Sep)	514
Lo. Indaal	106	342	-	336	561	(Aug)	336
Poole Hbr	387	168	338	389	380	(Nov)	332
Duddon Est.	262	281	271	291	431	(Oct)	307
Morecambe Bay	250	371	256	354	300	(Nov)	306
The Fleet/Wey	-	259	280	260	242	(Mar)	260
Cromarty Fth	332	+194	+340	+227	185	(Feb)	256
Tay Est.	427	216	223	23	65	(Oct)	191
Langstone Hbr.	234	185	186	111	153	(Dec)	174
Chichester Hbr	119	172	151	135	134	(Mar)	142
Clyde Est.	118	115	113	146	181	(Mar)	135
Exe Est.	154	112	92	158	111	(Dec)	125
Lindisfarne	198	132	81	(21)	32	(Nov/Mar)	111
Lo. Ryan	50	104	130	115	121	(Oct)	104
Wash	147	57	153	100	48	(Mar)	101
Northern Ireland							
Strangford Lo.	371	303	274	222	449	(Nov)	324
Belfast Lo.	181	234	204	142	173	(Feb)	187
Larne Lo.	181	218	125	207	167	(Oct)	180
Dundrum Bay	258	124	199	74	102	(Oct)	151
Lo. Foyle	42	42	48	17	84	(Nov)	47
Carlingford Lo.	45	31	40	30	35	(Feb)	36

+ RSPB/BP studies

GOOSANDER
Mergus merganser

				International importance:				1,500
				Great Britain importance:				90
				All-Ireland importance:				+*

GB maximum:	2,911	Jan	Trend	88-89	89-90	90-91	91-92	92-93
NI maximum:	1	Jan	GB	187	117	146	127	121

The peak count of British Goosander was 10% lower than those recorded in the last two seasons, whilst annual indices also show a decline since the late 1980s (Table 5). Numbers recorded represent only a small proportion of the British population, thought to be largely resident, though there is evidence of birds moving to the continent and vice versa, particularly in harsh weather conditions (Owen *et al.* 1986, Chandler 1986). Goosander remain a rarity in Northern Ireland, with only one bird recorded.

Owing to revised population estimates, many fewer sites hold nationally important numbers of Goosander than suggested previously (Table 38). Though the species is predominantly found inland, the key site remains a coastal one, namely the Inner Moray Firth, which until recent years regularly held in excess of 1,000 birds. Numbers at the site have, however, undergone a considerable decline since the late 1980s, with the 1992-93 peak count being exceeded by that at Hirsel Lake. Since Goosander disperse during the day to feed on rivers, evening counts of roosting birds would undoubtedly increase the apparent significance of specific sites. The small numbers of riverine sites in the WeBS scheme also contributes to the low proportion recorded. Other sites at which 90 or more birds were recorded during 1992-93 were Tyninghame Estuary (102, July) and Spey Mouth (95, August).

The Goosander population in Britain has risen steadily over the last three decades (Table 5), although annual indices indicate that the population undergoes cyclic fluctuations, with regular large troughs every four or five years (Kirby *et al.* in press, a). Since conditions in wintering areas are thought unlikely to give rise to such phenomena, this pattern may be related to ecological factors in the breeding areas, such as food availability.

Table 38. GOOSANDER: WINTER MAXIMA AT MAIN RESORTS

	88-89	89-90	90-91	91-92	92-93	(Mth)	Average
Great Britain							
Inner Moray Fth	1,490	273	+610	189	301	(Jan)	573
Hirsel Lake	124	202	290	170	360	(Oct)	229
Tay Est.	193	26	206	191	112	(Aug)	146
Hay-a-Park GP	53	195	166	165	137	(Dec)	143
Chew Valley Lake	107	110	163	114	91	(Dec)	117
Leighton/Roundhill Rsr	82	-	-	120	120	(Jan)	107
R. Tweed: Kelso/C'stream	91	88	147	109	75	(Mar)	102
R. Eden: R'cliffe/A'waite	60	111	110	-	-		94
Eccup Rsr	105	55	101	85	112	(Jan)	92

RUDDY DUCK
Oxyura jamaicensis

			International importance:	introduced
			Great Britain importance:	introduced++
			All-Ireland importance:	introduced

| GB maximum: | 2,253 | Dec |
| NI maximum: | 39 | Sep |

Trend	88-89	89-90	90-91	91-92	92-93
GB	5,737	6,148	6,103	6,341	4,884

As Ruddy Ducks in the UK are concentrated on relatively few wintering sites, the WeBS counts provide an accurate estimate of the number and distribution of birds. However, there are two notable absences in the winter maxima for Ruddy Ducks (Table 39). Firstly, Woolston Eyes has ceased to hold birds after the site was catastrophically drained in 1992 (Kirby & Owen 1993). This site not only held large numbers of wintering birds, but also the highest number of breeding Ruddy Ducks of any site in the UK: up to 30 pairs. Secondly, data from Chew Valley Lake was withheld; an unfortunate consequence of the present research into the distribution of and potential control measures for Ruddy Ducks in this country. The lack of these data precludes any meaningful comparison with the UK population in previous years, but it appears that, unless there was a record count at Chew in 1992-93, the number of birds in Great Britain has decreased for the first time in eight years (the apparent decline in 1991-92 was caused by missing data). Past decreases in numbers have probably been caused by cold weather mortality which, in the face of recent mild winters, cannot be used to explain the present decline. More likely, the apparent decline is due to reduced breeding success caused by summer weather conditions. Dry weather in June followed by a very wet July and August would have caused large fluctuations in water level on breeding sites favoured by this species, providing very poor conditions for breeding. Numbers in Northern Ireland remained relatively stable, peaking at around 40 birds in late autumn before birds dispersed away from WeBS sites during mid-winter.

Over the past two winters, the wintering distribution of Ruddy Ducks has shown a subtle change. The number of birds gathering on the two main West Midlands sites of Blithfield and Belvide Reservoirs has decreased considerably, while increased numbers have remained at more northerly sites, such as Swillington Ings in Yorkshire. The majority of birds now winter in the East Midlands and at the customary wintering site in Avon, Chew Valley Lake. The following sites also recorded 30 or more birds in 1992-93: Cropston Reservoir (70, December), Abberton Reservoir (52, March), Marton Mere (43, September), Hanmer Mere (41, September), Holme Pierrepont Gravel Pits (39, October) and Worsborough Reservoir (36, March).

Despite the apparent decline this winter, there is no reason to believe that the UK Ruddy Duck population will not continue to increase and with it the number of birds emigrating to the continent. The number of Ruddy Ducks wintering in continental countries continues to increase and birds are present during the breeding season in France, Belgium and the Netherlands, as well as in Spain where they continue to compete and hybridise with White-headed Ducks (e.g. Rose 1993). Spanish conservationists continue to eliminate all Ruddy Ducks and Ruddy Duck x White-headed Duck hybrids (16 Ruddy Ducks and 30 hybrids have been shot so far in Spain), but this job may become impossible if numbers in Europe continue to increase. To ensure the future existence of the White-headed Duck, the population expansion in Europe must be reversed. The present research programme in the UK will provide information on whether or not control methods are feasible and practical. Only when the results are known can any advice be given on the introduction of control on a wider scale.

Although much is already known about the wintering habits of Ruddy Ducks, much more information is needed on the highly dispersed breeding population and WWT will be carrying out a national breeding survey during 1994. This survey will count the number of adults present and broods reared at each site, and establish whether females can rear more than one brood per breeding season. Work is also underway to investigate movements of individual birds between sites during the winter, and birds caught at Rutland Water and Abberton Reservoir during 1993-94 will be fitted with white and yellow wing tags respectively which can be read using binoculars or a telescope.

Table 39. RUDDY DUCK: WINTER MAXIMA AT MAIN RESORTS

	88-89	89-90	90-91	91-92	92-93	(Mth)	Average
Great Britain							
Chew Valley Lake	785	470	435	760	-		612
Blithfield Rsr	640	508	899	402	394	(Dec)	569
Rutland Water	468	398	750	756	550	(Jan)	584
Eyebrook Rsr	218	302	304	352	101	(Dec)	255
Belvide Rsr	340	348	248	109	38	(Mar)	217
Swithland Rsr	55	81	184	192	284	(Mar)	159
Farmwood Pool	79	166	106	124	116	(Oct)	118
Stanford Rsr	57	195	37	110	190	(Dec)	118
Kingsbury/Coton Pools	57	85	133	106	132	(Oct)	103
Hanningfield Rsr	55	80	123	127	100	(Dec)	97
Hilfield Park Rsr	51	91	114	109	116	(Feb)	96
Blagdon Lake	173	61	108	71	58	(Dec)	94
Llyn Traffwll	81	97	84	100	81	(Oct)	89
Woolston Eyes	137	152	56	96	0		88
Staines Rsr	1	0	103	64	105	(Dec)	68
Llyn Penrhyn	71	118	101	-	46	(Feb)	67
New Swillington Ings	3	21	-	120	116	(Oct)	65
Alaw Rsr	54	57	43	52	85	(Feb)	58
Clumber Park Lake	5	49	33	53	73	(Mar)	43
Attenborough GP	7	17	81	55	36	(Dec)	39
Pitsford Rsr	4	4	23	27	120	(Feb)	36
Northern Ireland							
Lo. Neagh/Beg	44	19	30	38	37	(Sep)	34

++ since Ruddy Duck is an introduced species in the UK, a 1% criterion is not used for site designation. A threshold of 30 has been used here as the basis for selecting sites for presentation in this report.

WATER RAIL
Rallus aquaticus

International importance: ?
Great Britain importance: ?
All-Ireland importance: ?

GB maximum: 150 Oct
NI maximum: 7 Feb

Trend: not available

The wintering and breeding distribution of Water Rails are very similar. Although loose concentrations generally only occur during the winter period (Flegg 1986), most of the time the birds are solitary, and their secretive habits and skulking nature make them notoriously difficult to detect, other than by their squealing pig-like call. Despite their rather weak-looking flight, there is evidence from ringing recoveries that some continental birds over-winter in Britain, though there is little information on the movements of the British population. The peak total in Britain of 150 birds was far fewer than the 240 of 1991-92, although that count was influenced by the estimate of 100 birds for Stodmarsh. There was no clear pattern of numbers through the winter, fluctuating between 100 and 150 in most months (Table 1). The peak of only seven birds in Northern Ireland in 1992-93 compares with a maximum of five in the previous winter. A total of 11 sites held 10 or more birds, compared with only five in 1991-92: the Somerset Levels and Moors (27, December, cf. 24 in 1991-92), Longueville Marsh (20, November, not counted in 1991-92), Stodmarsh NNR (20, October, cf. 100), Lower Derwent Valley (20, March, not counted in 1991-92), Poole Harbour (15, October, cf. 7), Fleet Pond (15, January, cf. 10), Fairburn Ings (13, November, cf. 4), Llangorse Lake (12, December, cf. 4), Minsmere (11, February, cf. 5), Grouville Marsh (10, February, cf. 2) and Kenfig Pool (10, December, cf. 4).

MOORHEN
Gallinula chloropus

GB maximum:	7,738	Oct
NI maximum:	438	Oct

International importance: ?
Great Britain importance: ?
All-Ireland importance: ?

Trend   not available

Despite being one of the commonest waterfowl in Britain and Ireland, with a winter population estimated to be in excess of 1,000,000 birds (Taylor 1986), only a small fraction of the Moorhen population is counted by the WeBS scheme. The species' use of small waterbodies and rivers, combined with their bankside habits, make accurate counting difficult. Although, as in 1991-92, the peak occurred in late autumn, monthly totals in Britain showed little variation during 1992-93. Northern Ireland counts are more variable, although similarly, the peak has occurred in late autumn in both seasons during which Moorhen have been recorded by the WeBS scheme, with the 1991-92 total rather larger than that in 1992-93. The Moorhen's widespread distribution means few sites hold large concentrations. The following sites held 100 or more birds in 1992-93: Martin Mere (235, January, *cf.* 250 in 1992-93), Wantsum Marshes (195, January, not counted in 1991-92), Somerset Levels (166, March, *cf.* 135), Ash Levels (158, January, not counted in 1991-92), Little Stour Valley (150, January, not counted in 1991-92), Loughs Neagh and Beg (138, October, *cf.* 209), Blenheim Park Lake (124, October, *cf.* 104), Lancaster Canal (120, January, *cf.* 157), Chew Valley Lake (110, July, *cf.* 150), Fiddlers Ferry Power Station Lagoons (110, October, *cf.* 100), Washington WWT (104, October, *cf.* 125), the Ribble Estuary (103, January, *cf.* 20), the Ouse Washes (102, December, *cf.* 117) and Penshorpe Lakes (100, June to December, *cf.* 100).

COOT
Fulica atra

GB maximum:	95,319	Nov
NI maximum:	10,403	Dec

International importance: 15,000
Great Britain importance: 1,100
All-Ireland importance: 250

Trend	88-89	89-90	90-91	91-92	92-93
GB	111	114	100	98	101
NI	99	118	116	108	121

Coot are one of the commonest and most widely distributed waterfowl in the UK, their breeding and wintering areas being remarkably similar (Horsfall 1986). Peak winter numbers in 1992-93 were the highest recorded in Britain since 1989-90, and numbers in Northern Ireland exceeded 10,000 for the first time. Peak counts were unusually late in Northern Ireland (Table 3), perhaps because of the mild weather conditions.

Abberton Reservoir continues to hold by far the largest numbers of Coot in the UK, registering a large increase in numbers in 1992-93. Other British sites of note were Windermere and Tattershall Pits, both holding well above average numbers in 1992-93, while the Fleet/Wey maintained the increase numbers of recent seasons. Coot numbers at Rutland Water and Hanningfield, however, declined for a third consecutive year, while low counts were also made at Loch Leven and Little Paxton Gravel Pits. In Northern Ireland, the bulk of the birds were present at Loughs Neagh and Beg, which showed a considerable increase on previous seasons counts. Other sites which held 1,000 or more Coot in 1992-93 were Wraysbury Gravel Pits (1,240, January) and Hardley Flood (1,000, September).

Table 40. COOT: WINTER MAXIMA AT MAIN RESORTS

	88-89	89-90	90-91	91-92	92-93	(Mth)	Average
Great Britain							
Abberton Rsr	12,510	16,790	9,252	7,817	11,472	(Oct)	11,568
Rutland Water	4,160	5,502	3,743	3,639	2,785	(Jan)	3,966
Cotswold WP West	3,033	2,924	3,608	3,068	3,068	(Nov)	3,140
Hanningfield Rsr	1,983	4,350	3,668	2,870	2,105	(Aug)	2,995
The Fleet/Wey	(1,750)	1,561	2,647	3,489	3,344	(Dec)	2,760
Ouse Washes	1,537	2,345	-	1,940	2,289	(Nov)	2,028
Cotswold WP East	1,680	1,760	2,396	2,645	1,613	(Oct)	2,019
Avon Valley (Mid)	1,250	1,255	2,335	2,633	2,496	(Dec)	1,994
Stanford Rsr	1,750	2,115	1,118	1,800	1,000	(Sep)	1,557
Kingsbury/Coton Pools	1,936	1,459	1,310	1,296	1,338	(Dec)	1,468
Chew Valley Lake	1,055	1,290	1,070	1,945	1,950	(Aug)	1,462
Lo. Leven	2,270	1,630	1,515	1,250	430	(Sep)	1,419
Fen Drayton GP	1,112	2,090	950	1,300	-		1,363
Windermere	698	1,179	636	1,467	1,992	(Jan)	1,194
Cheddar Rsr	1,000	1,491	1,416	1,085	918	(Jan)	1,182
Chichester GP	1,011	1,118	1,099	1,584	910	(Dec)	1,144
Tattershall Pits	-	1,197	795	885	1,421	(Dec)	1,075
Fairburn Ings	1,511	1,275	1,151	564	715	(Sep)	1,043
Little Paxton GP	855	1,136	900	1,548	600	(Jan)	1,008
Thorpe WP	-	-	-	1,135	876	(Nov)	1,006
Northern Ireland							
Lo. Neagh/Beg	4,821	7,696	6,685	7,097	8,848	(Dec)	7,029
Strangford Lo.	763	359	274	222	595	(Dec)	443
Upper Lo. Erne	456	577	383	-	190	(Feb)	401
Ballysaggart Lo.	332	303	491	400	400	(Oct)	385

OYSTERCATCHER
Haematopus ostralegus

International importance: 9,000
Great Britain importance: 3,600
All-Ireland importance: 500

GB maximum: 291,469 Nov
NI maximum: 14,992 Feb

Trend	88-89	89-90	90-91	91-92	92-93
UK	161	157	167	153	165

In 1992-93 the UK winter index for Oystercatcher rose by 8%, almost reversing the 8% drop registered in the previous winter. Seven sites remain of international importance and these sites, together with those of national importance supporting average winter peaks above 5,000 birds and all sites important in an all-Ireland context are listed in Table 41. Other nationally important sites not listed in the table were the Eden, Exe, Inner Moray Firth, Lavan Sands, Medway, North Norfolk Marshes, Swale and Moray.

Concealed within the relatively small change in the overall UK winter index of 8% were two major sites registering 1992-93 winter peaks substantially different from the average of recent winters. At the Dee (Eng/Wales) the 1992-93 winter peak count of over 75,000 birds is more than double the average of the previous five years. In contrast the 1992-93 peak at the Wash was approximately 30% lower than the average of the preceding five year period. An examination of the counts on the Wash reveals that Oystercatcher numbers have been declining there over the past two winters. In addition to these low counts Oystercatcher corpses were found during the 1992-93 winter. The results of BTO research suggest that a shortage in the availability of mussels and cockles may explain the unusually high mortality (Clark 1993). It is predicted that within a couple of years the Wash will hold more shellfish of higher nutritional quality for Oystercatchers following good recent spatfall.

Oystercatchers have proved an excellent species for studying the effects of individual differences in feeding behaviour. Previous studies have described the effects of age and dominance but in a recent paper, Durell et al. (1993) describe some remarkable sex-related differences in the diet and feeding method of Oystercatchers on the Exe. More than 70% of the birds which fed on mudflats were females, whereas nearly 90% of the birds which hammered mussels were males. Birds feeding on the mudflats were found to have pointed bills, whilst mussel hammerers had blunt bills and stabbers had bills that were chisel shaped.

The Norwegian breeding population of Oystercatchers is one of the largest in Europe. Lambeck & Wessel (1993) recently reviewed the sparse ringing data available to examine the migration of Oystercatchers breeding in the high northern area of NE Norway. They suggested that birds breeding in coastal habitats west of North Cape - Porsanger migrate southwestwards along the Norwegian coast in autumn and winter chiefly in Britain. Birds breeding in East Finmark take a direct southwesterly flight of approximately 500 km over Lapland and along the coast of Bothnia to winter in an unknown region.

Table 41. OYSTERCATCHER: WINTER MAXIMA AT MAIN RESORTS

	88-89	89-90	90-91	91-92	92-93	(Mth)	Average
International							
Morecambe Bay	50,776	64,967	56,361	49,157	51,430	(Nov)	54,538
Dee (Eng/Wales)	27,397	33,293	35,774	35,681	75,778	(Jan)	41,584
Wash	46,912	40,689	33,791	(28,303)	26,618	(Nov)	37,002
Solway	28,536	(27,959)	40,095	36,533	(32,907)	(Nov)	35,054
Ribble	19,271	(15,046)	18,263	16,290	22,219	(Dec)	19,010
Burry	(14,980)	(11,862)	15,151	11,577	13,241	(Dec)	13,737
Thames	(8,297)	9,973	(17,378)	9,711	16,920	(Jan)	13,495
Great Britain+							
Forth	(7,600)	5,859	7,374	10,506	8,088	(Feb)	7,956
Carmarthen Bay	-	(750)	(3,779)	(3,309)	(7,334)	(Feb)	(7,334)
Duddon	6,401	8,428	5,898	5,880	8,620	(Nov)	7,045
Humber	(5,102)	(4,750)	(5,806)	(5,687)	(5,795)	(Jan)	(5,806)
Inner Clyde	3,679	4,914	4,348	6,215	5,962	(Feb)	5,023
Northern Ireland							
Belfast Lo.	5,594	4,480	5,601	5,787	7,583	(Jan)	5,809
Strangford Lo.	3,936	4,887	4,759	5,020	4,125	(Dec)	4,545
Lo. Foyle	2,550	(2,649)	2,599	1,845	1,683	(Dec)	2,265
Dundrum Bay	2,433	2,006	974	1,241	1,748	(Feb)	1,680
Outer Ards*	(1,473)	1,674	1,401	1,488	1,691	(Jan)	1,563
South Down*	-	1,502	-	-	-		1,502
Carlingford Lo.	858	659	870	650	(878)	(Jan)	783

+remaining sites of national importance listed in text

AVOCET
Recurvirostra avosetta

International importance: 700
Great Britain importance: 10*
All-Ireland importance: +*

GB maximum: 2,137 Feb
NI maximum: 0

Trend: not available

UK winter totals of Avocet continued to rise resulting in four of the winter months of 1992-93 recording their highest winter total. For the first time ever a UK site has attained international importance for wintering Avocet. Despite recording its lowest winter peak for three years the Alde complex is now recognised as internationally important for this species. All nationally important sites are listed in Table 42, along with the Alde complex. In addition to these sites the Wash (104 birds in March) recorded over 50 birds in the 1992-93 winter, although the five year average peak remains below the qualifying level. At these major sites winter peak counts of Avocet exceeded the recent averages except at the Alde and Tamar complexes, where the peak was slightly lower. As in previous years peak numbers on the Medway were in late autumn, on the Thames in mid-winter, whilst on the Swale numbers peaked in early spring. This provides strong circumstantial evidence that Avocet move between these three estuaries on a regular basis.

Reay (1993), in his continuing studies of Avocet wintering in SW England, examined the wintering season along different stretches of the south coast. It was noted that the length of the wintering period declined from east to west. A study by Hill & Player (1992) at Havergate Island found that methods of gull productivity control, whether by injecting gull eggs with formalin or raking nests and eggs, did not appear to affect Avocet behaviour.

Table 42. AVOCET: WINTER MAXIMA AT MAIN RESORTS

	88-89	89-90	90-91	91-92	92-93	(Mth)	Average
International							
Alde complex	514	721	729	946	(633)	(Feb)	727
Great Britain							
Exe	229	379	323	473	427	(Feb)	366
Hamford Water	85	(0)	188	227	298	(Nov)	199
Tamar complex	90	185	240	231	168	(Dec)	182
Poole Hbr	65	122	175	144	290	(Feb)	159
Medway	(38)	(136)	36	215	(188)	(Nov)	143
Thames	(40)	58	(37)	137	230	(Jan)	141
Swale	(18)	36	(75)	136	(94)	(Mar)	88
Blyth (Suffolk)	34	33	59	73	164	(Mar)	72
Deben	10	34	48	54	141	(Jan)	57
N Norfolk Marshes	-	13	110	93	56	(Mar)	54

RINGED PLOVER
Charadrius hiaticula

International importance:	500
Great Britain importance:	290
All-Ireland importance:	125

| GB maximum: | 10,512 | Nov |
| NI maximum: | 851 | Jan |

Trend	88-89	89-90	90-91	91-92	92-93
UK	133	123	113	103	104

The 1992-93 winter index for Ringed Plover rose just 1% on the previous year's value. This species appears to have one of the more stable winter populations amongst the waders with only one year-to-year index change exceeding 10% in the past ten years. All sites averaging over 400 birds and those sites important in an all-Ireland context are given in Table 43 which includes the seven sites of international importance. Only five of the 15 major sites recorded greater than average peaks in 1992-93, with the top two estuaries holding 25% and 50% fewer birds respectively.

As usual, many sites recorded high counts during passage periods, with 15 sites holding more than 500 birds in 1992-93. Over 1,000 birds were recorded at the Thames (2,106 in September), Chichester Harbour (1,482 in September), Severn (1,453 in August), Blackwater (1,229 in September), Humber (1,177 in August), Medway (1,116 in October), Wash (1,096 in August) and N Norfolk Marshes (1,054 in August). Spring passage was unusually light, even in NW England, where numbers recorded usually exceed those of the autumn. A count of 529 at Morecambe Bay in May was the only one above 500 throughout the whole country. As always it should be recognised that during passage periods turnover is much more rapid than in the winter, so that many more individuals are likely to have visited a site than is suggested by the number recorded on the monthly count dates.

Colonisation of inland sites, particularly in eastern and central England, has increased during the last 20 years (Gibbons et al. 1993), but still accounts for only 17% of the total breeding population in England. Prater (in Gibbons et al. 1993) estimates that the British and Irish total probably accounts for nearly 80% of the temperate breeding population of the nominate race (*C. h. hiaticula*) of Ringed Plover. This has particularly important conservation implications, especially as many Ringed Plovers which breed in western Europe also overwinter there. Any coastal habitat loss or degradation may obviously have serious consequences.

Table 43. RINGED PLOVER: WINTER MAXIMA AT MAIN RESORTS

	88-89	89-90	90-91	91-92	92-93	(Mth)	Average
International							
Thames	(605)	922	(674)	1,531	(803)	(Feb)	1,226
Chichester Hbr	924	(2,093)	519	630	744	(Dec)	982
Medway	(971)	(696)	501	581	770	(Dec)	703
Hamford Water	310	(0)	1,427	346	548	(Nov)	657
Outer Ards*	(753)	623	709	698	367	(Jan)	630
South Down*	-	529	-	-	-		529
Lindisfarne	720	311	800	(65)	215	(Feb)	511
Great Britain							
Stour	443	562	385	336	756	(Dec)	496
Morecambe Bay	497	440	380	435	568	(Mar)	464
Blackwater	564	613	160	(273)	(289)	(Dec)	445
Langstone Hbr	542	375	420	344	(274)	(Jan)	420
Solway	(580)	(391)	318	(340)	(232)	(Nov)	407
Northern Ireland							
Strangford Lo.	225	207	413	257	415	(Nov)	303
Belfast Lo.	169	236	179	85	215	(Dec)	176
Tyrella/Minerstown*	-	157	115	103	-		125

GOLDEN PLOVER
Pluvialis apricaria

International importance:	18,000
Great Britain importance:	2,500
All-Ireland importance:	2,000

GB maximum: 144,176 Dec
NI maximum: 19,769 Feb

Trend: not available

Estuarine/coastal UK totals of Golden Plover were above recent averages in all months of the 1992-93 winter. The short-lived peak of over 110,000 in December 1992 was comfortably the highest ever winter total. In contrast, in Northern Ireland the recorded monthly totals peaked at an all-time N Irish record of 15,034 in February. The very low counts of around 3,5000 in Northern Ireland in December and the record high UK counts in the same month may be the result of British birds being unwilling to disperse to other countries. At inland wetlands a pattern similar to that for the coast was evident, i.e. generally high numbers, especially in December, with Northern Ireland totals particularly low in the same month.

All sites of national and international importance, together with those Northern Irish sites with the highest average winter peaks, are listed in Table 44. There are now 18 sites of national importance, but for the first time the list includes six inland wetlands. The Humber remains our only site of international importance. As expected from the high overall UK totals, counts recorded at these major sites tended to be higher than recent averages. At most British sites the winter peaks were recorded in December. In addition to the nationally important sites, six estuaries recorded counts over 2,000 birds in the 1992-93 winter, although their five year average peak falls below this threshold.

Two recent papers have revealed contrasting fortunes in breeding populations of Golden Plover in two different parts of Britain. A survey of the south Pennines in 1990 (Brown 1993) revealed that within a study area of 725 km^2 there were an estimated 736 breeding pairs (overall density of 1.02 pairs/km^2). There was no evidence of any major change in abundance in recent decades in this southern breeding population. Work in NE Scotland, however, has shown that population declines have occurred in recent years, at least on some areas of moorland (Parr 1993). Experimental removal of predators (crows and gulls) did not lead to greater hatching success and Golden Plovers continued to decline.

Table 44. GOLDEN PLOVER: WINTER MAXIMA AT MAIN RESORTS

	88-89	89-90	90-91	91-92	92-93	(Mth)	Average
International							
Humber	(10,346)	(27,249)	(20,168)	25,946	16,969	(Dec)	23,388
Great Britain							
Lower Derwent#	-	-	-	-	6,800	(Dec)	6,800
Sutton/Lound GP#	-	-	-	-	6,200	(Dec)	6,200
Ribble	5,111	8,902	2,451	5,503	6,362	(Dec)	5,665
Solway	(4,360)	(3,706)	2,693	3,137	(12,321)	(Dec)	5,243
Netherfield GP#	-	-	-	3,000	6,000	(Dec)	4,500
Wash	3,568	2,248	(3,591)	(2,202)	7,657	(Dec)	4,491
Clifford GP#	-	-	-	3,000	5,000	(Dec)	4,000
Lindisfarne	4,000	3,200	1,600	(2,250)	6,050	(Dec)	3,712
Blackwater	1,551	2,213	3,388	961	(9,620)	(Feb)	3,546
Thames	(1,832)	(4,412)	(2,572)	2,063	(4,241)	(Feb)	3,322
Forth	1,742	(2,999)	6,637	2,007	2,012	(Dec)	3,099
Swale	(150)	(1,689)	(997)	1,968	(4,930)	(Mar)	2,862
Carmarthen Bay	-	(700)	0	(2,000)	(8,500)	(Jan)	2,800
Colliford Rsr#	-	-	-	2,000	3,400	(Nov)	2,700
Morecambe Bay	1,498	4,426	1,831	2,339	(3,128)	(Feb)	2,644
Northern Ireland							
Strangford Lo.	9,972	7,036	4,136	7,416	8,226	(Nov)	7,357
Lo. Neagh/Beg#	-	-	-	10,025	4,613	(Feb)	7,319
Lo. Foyle	600	(1,512)	2,828	5,095	5,700	(Feb)	3,555

GREY PLOVER
Pluvialis squatarola

International importance:	1,500
Great Britain importance:	440
All-Ireland importance:	40*

GB maximum:	38,465	Jan
NI maximum:	133	Jan

Trend	88-89	89-90	90-91	91-92	92-93
UK	459	414	462	478	442

The winter index for Grey Plover fell by 7% in 1992-93 compared to the previous winter. Sites with average winter peaks exceeding 1,000 birds and all sites important in an all-Ireland context are listed in Table 45. Of these 20 major sites, only three recorded a 1992-93 winter peak above the average of the previous five winters, with only Hamford Water registering a winter peak substantially greater than the recent average. Winter peak counts showing the most spectacular declines compared with the previous five year average were recorded at the Humber and Langstone Harbour. In addition to those sites listed in Table 45, Pagham Harbour and the Forth recorded counts greater than 1,000 in 1992-93, but their five year averages remain below this value. Other sites regularly holding nationally important numbers are the Beaulieu, Burry, Crouch/Roach, Duddon, Eden, Exe, Forth, Jersey*, North Norfolk Marshes, Orwell, Pagham Harbour, Pegwell Bay, Poole Harbour, Severn, Solway, Southampton Water, Taw/Torridge and Tay.

Grey Plover are affected by periods of severe winter weather, as shown by the significant reduction in their numbers on Suffolk estuaries in February 1992 (Wright in Piotrowski 1993). At least part of this decrease was attributed to cold weather mortality as shown by a 39% decline in the Deben's wintering population during the cold spell.

Pre-migratory studies of foraging Grey Plover in South Africa established that over 40% of their daily low-tide energy intake was achieved at night (Turpie & Hockey 1993). To compensate for their reduced visual acuity at night, Grey Plover foraged more slowly, paused for longer to search for prey and responded to closer visual cues than was apparent by day. The intake from nocturnal foraging supplements that from diurnal feeding and so enables Grey Plovers to accumulate sufficient reserves for migration.

Table 45. GREY PLOVER: WINTER MAXIMA AT MAIN RESORTS

	88-89	89-90	90-91	91-92	92-93	(Mth)	Average
International							
Wash	9,054	8,840	7,432	10,100	6,799	(Mar)	8,445
Thames	(8,486)	(4,835)	(6,388)	6,170	5,005	(Mar)	6,512
Medway	(3,209)	(6,185)	3,435	4,803	(4,619)	(Feb)	4,760
Ribble	3,539	(3,111)	2,720	4,148	2,655	(Jan)	3,265
Stour	1,761	2,473	1,999	4,279	3,152	(Feb)	2,732
Chichester Hbr	2,791	(1,591)	1,718	3,901	2,272	(Feb)	2,670
Humber	(1,425)	(1,343)	(2,490)	(1,338)	(1,123)	(Jan)	(2,490)
Blackwater	905	1,003	4,085	2,549	2,521	(Nov)	2,212
Swale	(1,362)	1,730	1,559	2,097	2,699	(Feb)	2,021
Dee (Eng/Wales)	1,270	1,120	2,004	3,420	1,471	(Nov)	1,857
Hamford Water	1,257	(3)	1,780	1,083	3,161	(Nov)	1,820
Morecambe Bay	3,062	1,074	1,466	1,714	1,747	(Nov)	1,812
Dengie	1,375	1,110	1,700	2,800	1,745	(Nov)	1,746
Great Britain+							
Langstone Hbr	1,870	1,196	1,299	1,682	997	(Dec)	1,408
Lindisfarne	1,825	1,200	1,020	(1,531)	1,334	(Feb)	1,382
Colne	1,063	1,540	1,083	1,294	944	(Dec)	1,184
Alt	1,276	1,340	990	1,051	(830)	(Mar)	1,164
Mersey	360	212	2,620	902	981	(Nov)	1,015
Northern Ireland							
Strangford Lo.	140	78	70	67	48	(Jan)	80
Outer Ards*	(33)	105	69	59	7	(Jan)	60

+ remaining sites of national importance listed in text

LAPWING
Vanellus vanellus

International importance:		20,000**
Great Britain importance:		20,000**
All-Ireland importance:		2,500

GB maximum: 519,355 Dec
NI maximum: 29,104 Jan

Trend not available

UK totals of Lapwing were well up on the recent averages in all months of the 1992-93 winter except January. The December coastal/estuarine total of over 387,000 was substantially higher than ever recorded in any previous winter month. British inland sites also recorded a peak total count in December. The counts for Northern Ireland were typical of recent winters and, in contrast to the UK totals, peak numbers were recorded in January.

Lapwing count totals show a similar pattern for the 1992-93 winter to those for Golden Plover, i.e. high British numbers with a huge peak in December, and a Northern Ireland peak in January. These two species are known to associate together and a large proportion of their UK wintering population occurs away from wetlands and is therefore not picked up by WeBS counts. Sites with average winter peaks above 5,000 birds plus those sites important in an all-Ireland context are listed in Table 46. It should be noted, however, that none of the Great Britain sites listed are of national importance. Only four sites recorded below average peaks in 1992-93 and two of these, the Humber and the Forth, were amongst the four major sites where below average counts of Golden Plover were also made in 1992-93. Exceptionally high counts of Lapwing were made at the Wash and the Blackwater in 1992-93, when winter peaks at both sites were more than 1,000% up on the average of the previous five winters.

Breeding populations of Lapwings on lowland grassland in England and Wales declined substantially in the 1980s. This was the conclusion of a survey of waders breeding on wet lowland grasslands in 1989 (O'Brien & Smith 1992). Between 1982 and 1989 there was a 38% decline in Lapwing breeding numbers in this habitat. Despite this decrease, Lapwing remains the most abundant breeding wader on lowland grassland.

Table 46. LAPWING: WINTER MAXIMA AT MAIN RESORTS

	88-89	89-90	90-91	91-92	92-93	(Mth)	Average
International							
Ribble	21,174	32,145	32,590	27,444	28,802	(Dec)	28,431
West Sedgemoor#	-	-	-	12,402	36,710	(Jan)	24,556
Wash	4,943	10,261	(3,007)	5,785	74,281	(Dec)	23,817
Humber	(12,644)	(30,892)	(26,506)	(15,009)	13,544	(Dec)	21,487
Morecambe Bay	26,327	24,171	12,247	18,857	23,309	(Dec)	20,982
Great Britain							
Lower Derwent#	-	-	-	-	11,700	(Mar)	11,700
Thames	(8,823)	(8,983)	(7,668)	9,643	13,659	(Mar)	11,651
Breydon Water	1,500	5,000	12,000	18,000	14,700	(Feb)	10,240
Pulborough Levels#	-	-	-	7,200	10,970	(Jan)	9,085
Blackwater	1,783	4,913	1,060	2,179	(32,700)	(Dec)	8,527
Mersey	2,512	6,150	11,720	12,500	8,584	(Jan)	8,293
Swale	(3,445)	(5,561)	4,915	8,202	11,113	(Dec)	8,076
Solway	(6,883)	(6,989)	3,504	9,697	(9,022)	(Dec)	7,219
Dee (Eng/Wales)	5,155	11,136	5,083	7,734	5,663	(Nov)	6,954
Crouch/Roach	(2,917)	(6,327)	3,257	5,053	12,534	(Dec)	6,948
Nene Washes#	-	-	-	3,581	10,150	(Feb)	6,865
Cotswold WP West#	-	-	-	(3,600)	6,470	(Dec)	6,470
Forth	4,091	5,124	7,529	5,953	4,898	(Dec)	5,519
Alde complex	5,205	4,635	3,393	2,766	9,840	(Dec)	5,167
Colne	4,611	5,361	1,908	2,074	11,114	(Dec)	5,013
Northern Ireland							
Strangford Lo.	13,592	11,826	10,651	9,074	9,591	(Mar)	10,946
Lo. Neagh/Beg#	-	-	-	13,501	5,595	(Dec)	9,548
Outer Ards*	6,492	5,688	3,915	9,070	8,280	(Jan)	6,689
Belfast Lo.	4,099	2,816	2,375	3,059	3,449	(Feb)	3,159

Knot
Calidris canutus

International importance: 3,500
Great Britain importance: 2,900
All-Ireland importance: 375

				Trend	88-89	89-90	90-91	91-92	92-93
GB maximum:	308,468	Dec		UK	112	104	106	110	113
NI maximum:	4,487	Feb							

The UK winter index for Knot changed by only 2% in 1992-93. Over the past five years the index has only varied between 103 and 113. Fifteen sites now qualify as nationally or internationally important for Knot and these are listed in Table 47. As would be expected in a year when the national index altered very little, about half the nationally and internationally important sites recorded above average peaks and about half recorded below average winter peaks. In addition to the sites listed in Table 47, Hamford Water held just over 3,500 Knot in January 1993 but counts over the previous four winters are not sufficient to push the average above 2,900.

Knot has one of the most spectacular migration systems of any bird species in the world and consequently has been studied extensively. However, there is still much to learn and Gudmundsson (1993) recently discovered new features about the spring migration of Knot through Iceland. Most birds tended to leave in the late afternoon when there was a tail wind taking them towards their high Arctic breeding grounds in Greenland and Canada. Many of us are used to seeing Knot in vast flocks but, when migrating, the average flock size was only 37 birds. Their flight direction suggested that the majority of birds were heading across the Greenland icecap toward northwest Greenland and Canada.

Identifying the food that is taken by waders is often difficult. In a recent study Dekinga & Piersma (1993) showed that it is possible to reconstruct the diet of Knot from examining faecal samples collected on the feeding grounds. By examining the minute fractions of shell in the faeces it was possible to identify not only the species of mollusc, but also its size and hence its food value to the Knot. By examining the rate of production of faeces in the field and comparing this with faecal analysis the authors were able to show a seasonal change in the birds diet from bivalves to snails with the approach of winter. Such seasonal shifts in diet reduce competition and allow estuaries to accommodate vast numbers of waders throughout the winter. Such studies help explain how our limited estuarine resource can accommodate vast numbers of waders throughout the winter.

Table 47. KNOT: WINTER MAXIMA AT MAIN RESORTS

	88-89	89-90	90-91	91-92	92-93	(Mth)	Average
International							
Wash	75,921	108,570	(164,176)	154,315	186,892	(Nov)	137,974
Ribble	60,030	(45,103)	30,567	42,644	54,400	(Dec)	46,910
Humber	(38,465)	(30,894)	(35,292)	(37,093)	(45,273)	(Jan)	(45,273)
Thames	(30,160)	(21,668)	(23,100)	35,650	(44,034)	(Dec)	39,842
Alt	45,000	51,000	28,000	20,001	(801)	(Mar)	36,000
Morecambe Bay	25,229	23,770	30,958	29,408	30,765	(Jan)	28,026
Dee (Eng/Wales)	13,132	44,715	16,916	21,016	37,700	(Jan)	26,695
Solway	(7,311)	(5,943)	15,305	11,604	7,652	(Feb)	11,520
Dengie	6,390	(7,300)	6,540	11,700	9,000	(Mar)	8,407
N Norfolk Marshes	6,260	(6,524)	13,298	6,142	6,822	(Jan)	8,130
Forth	10,810	7,744	7,163	6,743	5,730	(Jan)	7,638
Strangford Lo.	1,745	7,028	6,376	8,155	4,200	(Feb)	5,500
Swale	(3,503)	(2,101)	(3,208)	5,555	2,189	(Nov)	3,872
Duddon	600	2,300	5,570	2,743	8,000	(Jan)	3,842
Tees	4,484	3,000	1,953	3,403	5,938	(Jan)	3,755
Great Britain							
Montrose Basin	2,000	4,000	4,000	3,500	(3,100)	(Jan)	3,375

SANDERLING
Calidris alba

International importance: 1,000
Great Britain importance: 230
All-Ireland importance: 35*

GB maximum: 6,173 Dec
NI maximum: 85 Jan

Trend	88-89	89-90	90-91	91-92	92-93
UK	107	87	95	115	81

In 1991-92 the UK winter index for Sanderling rose to its highest level for 12 years. Between 1991-92 and 1992-93 the index dropped by 30% to reach its lowest level for 11 years. The index has only fallen significantly lower than the 1992-93 value in 1970-71. The number of Sanderling recorded by WeBS counts is approximately 40% of the UK wintering population, since the majority of the population overwinter on open coasts.

All nationally important sites with average peak counts over 250 birds are given in Table 48. As expected from the substantial drop in UK numbers, only one of those major sites recorded a 1992-93 winter peak count above its recent average. The largest declines were noted on the Wash (68%), Chichester Harbour (54%) and the Dee (Eng/Wales) (49%).

Passage UK totals were very low in 1992-93. In particular, the May total of 6,630 birds was less than half the average for the previous five years. High rates of turnover can mask the true numbers of birds using a site during migration periods, but such low counts give some cause for concern. Peak passage counts over 1,000 birds were recorded in 1992-93 at four sites: the Ribble (2,884 in August), Humber (1,392 in May), Morecambe Bay (1,326 in October) and Wash (1,153 in August). In typical years the highest counts tend to be recorded in May at sites in NW England. Thus 1992-93 was atypical both for the low wintering numbers recorded and for the unusual pattern during the migration periods.

Iceland is an important staging post for nearctic breeding Sanderling, thought to originate from northeast Greenland, particularly during spring migration. Observations in Iceland of colour-ringed individuals, marked at Teesside, have demonstrated the link with wintering grounds in western Europe, although ringing recoveries have come from as far south as Ghana (Gudmundsson & Lindstrom 1992). The Icelandic observations also indicated that individuals display a high stopover fidelity between years. Stopover durations of up to 15 days have been recorded, with most Sanderling arriving in Iceland after 10 May and departure being concentrated in the period 25 May to 4 June.

Table 48. SANDERLING: WINTER MAXIMA AT MAIN RESORTS

	88-89	89-90	90-91	91-92	92-93	(Mth)	Average
International							
Ribble	3,574	(1,460)	2,200	2,856	2,744	(Dec)	2,843
Great Britain							
Dee (Eng/Wales)	186	823	1,011	1,581	419	(Nov)	804
Humber	(472)	(556)	(559)	761	(537)	(Feb)	761
Carmarthen Bay	-	(0)	(484)	(654)	(218)	(Jan)	(654)
Thanet*	572	604	566	610	(488)	(Feb)	588
Hartlepool-Sunderland*	-	-	-	525	(292)	(Mar)	525
Alt	429	680	488	391	(320)	(Mar)	497
Duddon	388	457	383	600	425	(Feb)	450
Wash	435	471	302	378	140	(Jan)	345
Tay	336	312	380	300	180	(Nov)	301
Chichester Hbr	300	(432)	253	325	149	(Jan)	291
N Norfolk Marshes	60	311	371	339	320	(Nov)	280
Jersey*	228	285	346	260	277	(Jan)	279
Morecambe Bay	85	171	414	494	164	(Mar)	265
Thames	(33)	(11)	(120)	(262)	(185)	(Nov)	(262)

LITTLE STINT
Calidris minuta

GB maximum:	4	Nov
NI maximum:	0	

Winter counts of more than two Little Stints at any one site are rare in the UK and during the 1992-93 winter no sites recorded more than two birds. Passage numbers, normally higher than those recorded in winter, were below recent averages with eight sites holding more than five birds (all autumn counts). Peak passage counts were on the Thames (10 in August), Wash (eight in September), Crouch/Roach (eight in September), Mersey (seven in September), North Norfolk Marshes (seven in September), Southampton Water (six in September), Humber (six in October) and, perhaps most interestingly Ythan (six in September).

Recent notes by Chylarecki & Kania (1992a, 1992b) describing observations of colour-ringed Little Stints in Taimyr, Siberia proved evidence of polygyny and polyandry in the species as well as producing a time budget of incubation and chick-rearing.

Serra et al. (1992) made weekly counts of Little Stint at Saline di Cervia NE Italy between February 1990 and May 1991. Counts were compared with capture data to provide a detailed phenological description of migrant and wintering populations. Peak counts were recorded during pre-breeding migration (May) with slightly lower figures during post-breeding migration (August-October). Significant differences in body mass were observed amongst captured birds according to age, moult condition and time of year.

CURLEW SANDPIPER
Calidris ferruginea

GB maximum:	1	Nov
NI maximum:	0	

The Curlew Sandpiper on the Wash in November 1992 provided the only winter record for this species which is far more numerous on passage, especially in the autumn. UK totals in autumn 1992 were well below recent averages and only one count at one site exceeded 25, with 55 birds present on the Thames in August. The previous autumn had produced far more records of Curlew Sandpipers, when 10 sites held more than 25 birds.

Interestingly, a juvenile ringed near Giske, Norway on 30-8-92 was recovered at Harty, Isle of Sheppey on 25-9-92, having covered a distance of 1,277 km.

Martin et al. (1992) have published the first known movement of a Curlew Sandpiper between southern Africa and the Arabian Gulf. On examination the bird was shown to have modified its moult in line with other birds present in the Gulf.

PURPLE SANDPIPER
Calidris maritima

International importance:		500
Great Britain importance:		210
All-Ireland importance:		10*

GB maximum:	1,698	Mar	
NI maximum:	87	Jan	

Trend   not available

UK totals of Purple Sandpiper recorded in the 1992-93 winter were greater than those of the previous year and close to the average values of the previous five years. Winter counts over 100 birds were recorded at the following sites in 1992-93: Hartlepool-Sunderland* (347 in February), Rosehearty-Fraserburgh* (346 in March), Newbiggin-Blyth* (313 in March), Seahouses-Budle* (278 in February), Tees (185 in March), Dee (Scotland) (166 in November), Lossiemouth* (130 in February) and the Spey Coast* (110 in January). As usual, all these major sites are in NE Britain and in 1992-93 only two recorded a winter peak count below recent averages.

A wintering site is recognised as nationally important for Purple Sandpiper if the peak counts average more than 210 birds. Hartlepool-Sunderland*, Rosehearty-Fraserburgh* and the Tees currently qualify as nationally important. It is particularly pleasing to find the Hartlepool-Sunderland stretch of open coast at the top of the list, since the newly installed local organiser and his counting team have worked hard to cover this important stretch for the first time.

A study of seasonal patterns in the body mass of Purple Sandpipers wintering in Britain failed to detect significant variation in mass during the winter (Summers *et al.* 1992). This is in contrast to the patterns found with other waders such as Dunlin, Sanderling and Turnstone which all showed a significant increase in mass in mid-winter. It was suggested that Purple Sandpipers may be less at risk from cold weather and food shortages, perhaps because their food supply on rocky shores is less affected by cold weather than the prey of other wader species in estuaries. The Purple Sandpipers in this study displayed an increase in mass in May prior to migration, with those birds travelling the furthest distances accumulating the greatest fat reserves.

DUNLIN
Calidris alpina

International importance:	14,000
Great Britain importance:	5,300
All-Ireland importance:	1,250

GB maximum:	448,485	Dec
NI maximum:	11,935	Jan

Trend	88-89	89-90	90-91	91-92	92-93
UK	83	85	100	97	83

Dunlin is the only species where the UK winter index has not risen above 100 for more than 15 years. In 1992-93 the index dropped 15% on the previous winter, to reach its lowest level for five years. All internationally important sites, plus those of national importance that now average over 10,000 birds and all sites important in an all-Ireland context are listed in Table 49. Additionally, the Burry, Dengie, Duddon, Exe, Forth, Hamford Water, Lindisfarne, Moray Firth, Orwell, Poole Harbour, Southampton Water and Tamar support nationally important numbers of Dunlin. At the top six sites the winter peak was well down on recent averages but surprisingly the majority of the remaining sites in this table registered above average peaks, although the only notably high count was at the Blackwater.

Many Dunlin coming to Britain migrate through the Baltic and a recent paper showed that some of these birds actively moult their wing feathers before they arrive in W Europe (Holmgren *et al.* 1993). Second year birds were more likely to be in moult than older birds and tended to stop for longer periods on migration in the Baltic than older birds but could gain weight at the same rate. Those birds that arrived with low body weights tended to gain weight more rapidly than those arriving with some fat already laid down.

During the year, the BTO carried out a research project looking at the amount of movement between roost sites on the Wash (Rehfisch *et al.* 1993). The study used the wealth of information gathered by the Wash Wader Ringing Group over the last 30 years. Juvenile Dunlin were found to be rather more mobile than adults, both within winters and from year to year. Nevertheless, over three quarters of juveniles were retrapped in the same part of the Wash as they were ringed. Only 10% of adults were recorded in different parts of the Wash within the same non-breeding season and, remarkably, a similar proportion were found to move over several years. This indicates that most Dunlin return to the same part of the Wash each year.

Table 49. DUNLIN: WINTER MAXIMA AT MAIN RESORTS

	88-89	89-90	90-91	91-92	92-93	(Mth)	Average
International							
Morecambe Bay	42,987	54,802	76,602	72,113	49,285	(Jan)	59,157
Severn	44,311	(44,170)	58,705	42,056	(35,611)	(Feb)	48,357
Wash	65,679	56,510	43,233	43,768	29,680	(Jan)	47,774
Mersey	22,000	17,500	52,100	55,000	30,000	(Dec)	35,320
Langstone Hbr	31,700	37,660	27,720	34,500	31,250	(Feb)	32,566
Thames	(24,309)	25,893	29,925	38,556	(22,968)	(Feb)	31,458
Medway	(28,569)	(21,843)	26,442	28,607	(29,753)	(Feb)	28,342
Ribble	16,684	(14,147)	19,038	39,832	30,862	(Jan)	26,604
Humber	(21,899)	(22,903)	(26,133)	25,604	(27,075)	(Jan)	26,270
Dee (Eng/Wales)	16,772	14,710	24,670	31,368	21,223	(Dec)	21,748
Chichester Hbr	12,915	(28,268)	24,235	13,972	21,721	(Dec)	20,222
Blackwater	19,785	11,400	19,025	20,900	(26,425)	(Dec)	19,507
Stour	16,154	16,116	16,429	17,412	19,902	(Nov)	17,202
Great Britain+							
Solway	(12,443)	(14,537)	12,977	14,404	9,572	(Mar)	12,786
Swale	(13,610)	(12,055)	12,410	11,785	12,988	(Dec)	12,698
Colne	10,933	12,930	12,506	12,092	10,190	(Jan)	11,730
Northern Ireland							
Strangford Lo.	5,128	10,693	5,043	5,010	2,403	(Jan)	5,655
Lo. Foyle	1,900	(2,900)	2,475	(3,000)	5,170	(Feb)	3,181
Outer Ards*	2,967	2,506	2,620	2,493	1,744	(Jan)	2,466
Carlingford Lo.	3,300	910	1,420	1,670	(688)	(Jan)	1,825
Belfast Lo.	997	1,263	1,376	2,416	1,659	(Dec)	1,542
Dundrum Bay	1,550	(2,100)	1,991	1,345	465	(Jan)	1,490

+ remaining sites of national importance listed in text.

RUFF
Philomachus pugnax

International importance: ?
Great Britain importance: 7*
All-Ireland importance: +*

GB maximum: 420 Mar   Trend   not available
NI maximum: 3 Nov

In the 1991-92 winter totals of Ruff recorded at estuarine/coastal sites in the UK reached their highest levels for more than five years. Numbers recorded in 1992-93 were more typical of recent winters. In contrast UK totals on inland wetlands were higher in the 1992-93 winter than the previous year, although differences in coverage may be partly responsible. Winter peak counts greater than 20 were recorded at 10 sites, mostly in SE England: Lower Derwent# (107 in March), the Blackwater (63 in February), Martin Mere# (57 in March), Nene Washes# (46 in January), Hamford Water (45 in December), North Norfolk Marshes (45 in March), the Swale (45 in December), Pulborough Levels# (35 in January), Sandbach Flashes# (35 in January) and the Crouch/Roach (25 in November). Passage counts greater than 50 birds were recorded at Cresswell-Chevington* (140 in September), Humber (106 in September) and North Norfolk Marshes (56 in September).

The numbers of Ruff wintering in the UK are small compared with recent estimates of 180-200,000 birds in the Senegalese part of the Senegal River delta in January (Trolliet et al. 1992). In Germany migrant Ruffs were censused in both July and August 1990, (OAG Münster & OAG Schleswig-Holstein 1992). Around 5,000 birds were counted on each date, mostly at the coast, but the high proportion of juveniles in August suggested high turnover. Papers collated by Hötker (1991) report decreases in Ruff as a breeding bird in The Netherlands, Belgium, Germany and Denmark.

The highly complex lekking behaviour and mating system of Ruffs has been recently described by Hogan-Warburg (1992), and Hoglund et al. (1993). Hogan-Warburg draws attention to the similarities in location, configuration and spacing of individuals, between the lek and groups of birds feeding on spring passage. The author relates this to van Rhijn's (1991) hypothesis that many females are inseminated at leks encountered during their spring migration. It is suggested that the intense competition between males for mating opportunities has led to the evolution of a system in which passing females are intercepted at foraging stops far south of their eventual nesting location.

Jack Snipe
Lymnocryptes minimus

GB maximum:	85	Nov		
NI maximum:	1	Jan		

International importance: ?
Great Britain importance: ?
All-Ireland importance: 250

Trend not available

UK totals for Jack Snipe in the 1992-93 winter were close to recent averages for estuarine/coastal sites, but for inland wetlands UK totals were significantly higher than the previous year. The cryptic plumage and skulking habits of this species mean that the counts probably record only a small fraction of the birds present. During the 1992-93 winter double figures were recorded on only one occasion when 10 were seen at Chichester Harbour in February.

SNIPE
Gallingo gallingo

GB maximum:	6,729	Nov
NI maximum:	89	Feb

International importance: 10,000
Great Britain importance: ?
All-Ireland importance: ?

Trend not available

The total numbers of Snipe recorded at estuarine/coastal sites in the UK during the 1992-93 winter were again generally lower than recent averages, particularly in January. As in 1991-92, UK totals recorded at inland sites were somewhat higher than those at estuarine/coastal sites. For only the second time in more than 10 years, and for the second consecutive winter, no estuarine/coastal sites recorded a winter count above 200 birds. This figure was exceeded at six inland sites: Stodmarsh# (1,000 in November), W. Sedgemoor# (605 in February), Burtle Moor# (473 birds in December), Pulborough Levels# (405 in November), Tealham and Tadham Moors# (351 in February) and Nene Washes# (251 in November).

The breeding origins of the UK wintering Snipe are largely in Scandinavia and the UK. Surveys conducted in 1982 and 1989 showed that a high percentage of British breeding Snipe are now restricted to nature reserves, where their numbers have declined less than on unprotected areas (O'Brien & Smith 1992). In Britain and Ireland, a contraction of breeding range has occurred in recent years, especially in lowland regions of S England, Wales, NE Scotland and S Eire. The British and Irish breeding population now stands at c. 45,000 pairs having declined by approximately 20% since 1972 (Gibbons et al. 1993). This decline is thought to have resulted from the drainage of suitable wetland nesting habitats for intensive agricultural purposes. Undrained areas would normally remain waterlogged throughout most of the summer thereby prolonging the availability of soil invertebrates for Snipe offspring.

By examining bags of Snipe shot in the UK over the past 100 years, Tapper (1992) showed that wintering numbers were almost certainly much greater in the first half of the twentieth century than in the past forty years. In contrast to recent breeding surveys, however, winter bag returns give no evidence of any continuing decline over the past 30 years. It is suggested that this pattern may be the result of changes in agricultural practice, with the heaviest declines caused by periods of extensive land drainage. A decrease in the wintering population is also suggested in Denmark where shooting bags have declined by a half since the 1950s (Meltofte 1993). By comparing recoveries of hunted and non-hunted ringed Snipe across Europe, Henderson et al. (1993) showed that hunting pressure is unlikely to be the cause of the declines in the European Snipe population. Instead habitat loss was thought to be the more likely cause.

BLACK-TAILED GODWIT
Limosa limosa

	International importance:	700
	Great Britain importance:	75
	All-Ireland importance:	90

GB maximum: 9,940 Nov
NI maximum: 440 Jan

Trend	88-89	89-90	90-91	91-92	92-93
UK	167	188	143	168	183

The 1992-93 UK winter index for Black-tailed Godwit continued its upward trend with a 9% increase over the previous winter. All internationally important sites, all sites important in an all-Ireland context plus those of national importance that average over 400 birds are presented in Table 50. The Alde, Beaulieu, Blyth, Suffolk, Burry, Cefni, Crouch/Roach, Deben, Dengie, Eden, Fal, Humber, Mersey, Morecambe Bay, Newtown, Thames, North West Solent, Portsmouth Harbour, Solway and Tay are also of national importance for Black-tailed Godwit. At about half of these major sites, the winter peak for 1992-93 was below recent averages. The Colne and, particularly, the Ribble held substantially fewer birds than in the 1991-92 winter, when numbers had also declined from previous years. Low recorded counts on the Ribble in 1992-93 may be partly the result of incomplete coverage achieved, but it is highly probable that actual numbers present were indeed very low that winter. Typically, UK monthly totals during passage periods were higher than those in the winter, and with more rapid turnover occurring during the migration periods, far more birds must pass through the UK during spring and autumn than overwinter. Passage counts greater than 700 were made at nine sites in 1992-93, compared to 12 in the previous year. Peak passage counts of over 1,000 birds were made at Hamford Water (3,058 in September), Dee (Eng/Wales) (1,901 in October), Ribble (1,586 in August), Medway (1,549 in August), Stour (1,090 in October) and Wash (1,062 in April).

The number of Black-tailed Godwits occurring on the Wash has increased dramatically since regular counting for the BoEE began in 1969. The results of an analysis of the biometric data of all Black-tailed Godwits ringed by the Wash Wader Ringing Group since 1970 (Clark *et al.* 1993) supported the suggestion that the birds were from the *islandica* population (Prater 1981). On the Ouse Washes, the breeding population considered to be an extension of the Dutch *limosa* population has suffered more mixed fortunes (Gibbons *et al.* 1993). During 1988, 1989 and 1990, 33-36 pairs were confirmed to have bred in Britain, compared with approximately 60 pairs in the late 1970s.

Table 50. BLACK-TAILED GODWIT: WINTER MAXIMA AT MAIN RESORTS

	88-89	89-90	90-91	91-92	92-93	(Mth)	Average
International							
Swale	465	(34)	(569)	2,115	1,394	(Feb)	1,754
Stour	1,080	(1,734)	2,372	2,169	1,007	(Feb)	1,672
Hamford Water	1,010	(70)	2,241	1,254	1,899	(Nov)	1,601
Dee (Eng/Wales)	552	1,600	1,233	1,617	1,760	(Nov)	1,352
Poole Hbr	1,099	1,451	1,236	1,280	1,423	(Mar)	1,297
Ribble	2,490	491	977	561	(38)	(Mar)	1,129
Blackwater	392	(1,037)	743	1,132	1,167	(Mar)	894
Great Britain+							
Chichester Hbr	1,125	(750)	367	536	451	(Jan)	645
Colne	1,400	616	378	147	499	(Mar)	608
Southampton Water	427	(997)	311	305	876	(Dec)	583
Exe	542	648	782	480	450	(Jan)	580
Langstone Hbr	761	599	651	460	305	(Dec)	555
Wash	132	664	(401)	321	854	(Dec)	492
Medway	(519)	(630)	168	274	856	(Mar)	489
Orwell	139	(335)	330	700	597	(Dec)	441
Northern Ireland							
Belfast Lo.	138	135	139	286	330	(Jan)	205
Strangford Lo.	176	175	47	121	110	(Jan)	125

+ *remaining sites of national importance listed in text.*

BAR-TAILED GODWIT
Limosa lapponica

				International importance:	1,000
				Great Britain importance:	500
				All-Ireland importance:	175

GB maximum:	40,778	Nov	Trend	88-89	89-90	90-91	91-92	92-93
NI maximum:	2,885	Jan	UK	129	126	152	122	120

Although the UK winter index for Bar-tailed Godwit fell by only 2% in 1992-93, the new value is the lowest for 12 years. Table 51 lists all sites of international importance, those sites important in an all-Ireland context and those of national importance averaging over 600 birds. Other nationally important sites not listed in the table are the Inner Moray Firth and the North Norfolk Marshes. Winter peaks were below the recent average for almost all sites in the top half of the table but all bar two sites in the bottom half recorded a 1992-93 winter peak that was above the recent average. For a species whose counts fluctuate widely (often due to large scale international movements), 1992-93 was an unremarkable winter at all major sites except the Solway, where the winter peak of 1,940 birds was only 48% of the average peak count over the previous five winters.

Larsen (1993) has studied the foraging behaviour of Bar-tailed Godwits on their breeding grounds in Norway in the period prior to egg-laying. The study area consisted of a mosaic of peat bogs, shallow lakes and ponds, and dry, lichen covered moraines with open birch forests. Female godwits preferred to feed in the wet bogs, whilst males associated with Whimbrel foraging on the dry mounds (palsas) in the bogs. The vigilance of male godwits was not significantly affected by the presence of their mates, but female vigilance fell significantly both when they fed near their mate and when they fed near Whimbrels. Female godwits therefore, share their vigilance with Whimbrels whereas male godwits do not appear to benefit from interspecific association.

Table 51. BAR-TAILED GODWIT: WINTER MAXIMA AT MAIN RESORTS

	88-89	89-90	90-91	91-92	92-93	(Mth)	Average
International							
Ribble	7,898	13,350	9,940	18,775	(10,412)	(Mar)	12,490
Wash	8,403	12,622	14,834	9,807	11,098	(Dec)	11,352
Thames	(3,304)	(3,804)	(11,517)	3,969	(9,530)	(Nov)	8,338
Alt	7,902	5,391	7,095	2,934	(3,913)	(Mar)	5,830
Lindisfarne	6,010	6,200	4,900	(3,590)	(3,515)	(Jan)	5,703
Solway	7,315	(2,831)	3,650	1,536	(1,940)	(Dec)	4,167
Forth	3,372	1,510	2,722	3,075	2,260	(Feb)	2,587
Lo. Foyle	2,520	(2,222)	3,427	1,115	2,140	(Jan)	2,300
Humber	(1,054)	(1,270)	(2,002)	(1,837)	(1,711)	(Jan)	(2,002)
Morecambe Bay	1,844	858	2,568	1,886	2,736	(Nov)	1,978
Inner Moray Fth	1,465	1,487	1,987	1,632	2,374	(Feb)	1,789
Tay	1,835	1,400	1,696	2,296	1,310	(Jan)	1,707
N Norfolk Marshes	423	(1,599)	1,653	1,225	852	(Jan)	1,150
Chichester Hbr	890	(1,448)	1,056	954	1,267	(Dec)	1,123
Cromarty Fth	907	801	1,309	913	1,231	(Jan)	1,032
Dee (Eng/Wales)	152	396	2,480	837	(1,181)	(Feb)	1,009
Great Britain[+]							
Dornoch Fth	633	546	1,515	995	1,050	(Feb)	947
Dengie	386	800	1,000	1,200	1,180	(Mar)	913
Eden	892	700	680	490	1,461	(Feb)	844
Northern Ireland							
Strangford Lo.	1,074	628	329	291	836	(Mar)	631

+ *remaining sites of national importance listed in text.*

WHIMBREL
Numenius phaeopus

			International importance:	6,500
			Great Britain importance:	+*
			All-Ireland importance:	+*

| GB maximum: | 20 | Jan |
| NI maximum: | 3 | Mar |

UK totals for Whimbrel were typically in single figures for all months of the 1992-93 winter except in January, when a count of 15 at Artro was largely responsible for the UK total of 20 birds. Peak passage counts were, as usual, much higher with peaks over 100 recorded at the Severn (254 in May), Carmarthen Bay (229 in October), Thames (189 in April), Wash (177 in July), Morecambe Bay (147 in May), Lough Foyle (146 in May), Chichester Harbour (132 in May) and Humber (105 in May).

The nocturnal foraging of Whimbrel was studied on the Zwartkops estuary, South Africa by Turpie & Hockey (1993). Whimbrels foraged more slowly at night (to compensate for the reduced visibility) but their energy intake rates did not differ significantly from those achieved during the day. A long-term study of breeding waders in Finland has shown that breeding densities of Whimbrel were appreciably higher on clear-cut areas of forestry than in natural open habitats (Pulliainen & Saari 1993). Extensive clear-cut areas in Lapland have provided new breeding grounds for Whimbrel, but the population nonetheless seems to be declining, suggesting that the availability of nesting habitat is not limiting.

A recent recovery of an adult ringed on Fetlar confirmed the wintering area of at least some of our breeding birds. This individual was controlled in November at Guinea-Bissau, four years after being ringed on its nest in Scotland.

CURLEW
Numenius arquata

			International importance:			3,500
			Great Britain importance:			1,200
			All-Ireland importance:			875

| GB maximum: | 85,546 | Dec |
| NI maximum: | 7,744 | Dec |

Trend	88-89	89-90	90-91	91-92	92-93
UK	126	135	123	148	151

In 1992-93 the UK winter index for Curlew rose by just 2% on the value of the previous year. This is the highest value recorded since counting began in 1970. As in the case of the Lapwing and Golden Plover (that also winter in the UK in significant numbers away from wetlands) the highest UK totals for 1992-93 were recorded in December. In contrast to the two plover species, winter count totals of Curlew peaked in Northern Ireland in December. All internationally important sites, together with those of national importance averaging over 1,750 birds and all sites important in an all-Ireland context are listed in Table 52. In addition, the Burry, Chichester Harbour, Cleddau, Clyde, Cromarty Firth, Dornoch Firth, Lavan Sands, Mersey, Poole Harbour, Stour, Swale, Taw/Torridge and Wigtown Bay also hold nationally important numbers of Curlew. At the majority of these sites the peak counts for the 1992-93 winter were above the recent averages.

Territory establishment and habitat use by breeding Curlew have been examined in a recent study in central Sweden (Berg 1992, 1993). Food availability was thought to be important in the establishment of territories. Earthworms were found to be the most important food item during this period. Significantly higher numbers of earthworms were caught per minute in sown grassland than in tillage, despite there being no significant difference in the biomass of earthworms between the two habitats. Breeding densities of Curlew were higher on grassland than farmland where tillage predominates because of greater food availability. These results suggest that changes in land use due to intensification of farming practices have caused the Swedish Curlew population to decline since the 1950s.

Table 52. CURLEW: WINTER MAXIMA AT MAIN RESORTS

	88-89	89-90	90-91	91-92	92-93	(Mth)	Average
International							
Morecambe Bay	9,849	10,199	13,174	12,970	14,538	(Dec)	12,146
Solway	(3,757)	(4,882)	5,171	7,360	(5,345)	(Dec)	6,265
Wash	3,796	3,295	3,578	(3,727)	4,396	(Mar)	3,766
Dee (Eng/Wales)	2,474	2,910	2,892	5,331	4,209	(Jan)	3,563
Great Britain+							
Severn	2,706	(2,736)	2,505	(3,328)	4,555	(Dec)	3,273
Thames	(3,492)	(3,345)	(3,301)	3,311	2,380	(Mar)	3,165
Humber	(2,704)	(1,483)	(2,320)	3,414	2,344	(Dec)	2,879
Forth	(1,306)	1,676	2,137	2,520	2,859	(Feb)	2,298
Duddon	2,163	2,300	1,992	2,094	2,342	(Dec)	2,178
Inner Moray Fth	1,355	1,929	2,293	(1,726)	2,491	(Feb)	2,017
Orkney (Widewall)*	1,750	2,400	1,700	2,200	-		2,012
Blackwater	1,067	2,102	2,401	2,706	1,622	(Mar)	1,979
Medway	(1,796)	(1,981)	1,868	1,986	1,932	(Dec)	1,941
Northern Ireland							
Lo. Foyle	3,000	(1,351)	1,925	1,982	2,439	(Dec)	2,336
Strangford Lo.	2,056	1,483	2,096	1,575	2,467	(Dec)	1,935
Outer Ards*	1,501	1,997	710	1,793	758	(Jan)	1,351
Belfast Lo.	1,377	962	1,128	1,440	1,096	(Nov)	1,200

+ remaining sites of national importance listed in text.

SPOTTED REDSHANK
Tringa erythropus

GB maximum:	70 Nov
NI maximum:	2 Nov/Feb

International importance: 1,500
Great Britain importance: +*
All-Ireland importance: +*

Trend: not available

The recorded UK totals of Spotted Redshank were typically around 50 birds each month during the 1992-93 winter. Winter counts in double figures were recorded at the Tamar complex (20 in November) and on the Medway (11 in December). Passage period UK totals were unexceptional and peak counts over 50 were recorded only at the Wash (128 in July), Swale (74 in July) and the Blyth (Suffolk) (61 in September). Both these sites also recorded passage peaks above 50 in 1991-92.

The amount of food removed by several wader species including Spotted Redshank was assessed by Sezkely & Bamber (1992) by excluding birds from certain plots which contained predominately Chironomid larvae. These experiments in central Europe were backed up by direct observations of Black-tailed Godwit, Ruff and Spotted Redshank. Over the 13 day experiment, around 85% of the prey were removed by the waders present.

REDSHANK
Tringa totanus

			International importance:				1,500
			Great Britain importance:				1,100
			All-Ireland importance:				245

GB maximum:	73,606	Dec	Trend	88-89	89-90	90-91	91-92	92-93
NI maximum:	6,358	Feb	UK	119	120	101	112	103

In 1992-93 the UK winter index for Redshank fell by 8% on the previous year. UK totals were close to recent averages in all months of the 1992-93 winter. Table 53 lists all internationally important sites, together with those of national importance and all sites of importance in an all-Ireland context. About half of the major sites recorded peak counts above the recent averages but only at the Alde complex, Poole Harbour and Blackwater were these peaks counts up more than 50%. Only at Lindisfarne was the peak less than half of the recent average.

Redshank can suffer high mortality at winter sites from raptor predation. Escape responses have been shown to depend on the species of predator and the type of attack (Cresswell 1993). Behaviour that increases the chance of identifying an approaching predator may elicit an appropriate escape response.

Long-term studies at a coastal breeding site found that there was no significant difference in male and female adult survival rates each year (Thompson & Hale 1993). There were, however, consistently more females in the study population, with females breeding more frequently in their first year than males. Breeding numbers were lower, though not significantly so, in years following a wet June and were higher following a wet July. The number of breeding males, but not females, was found to be partially related to the mean February air temperature in the previous winter, with fewer males returning to breed following a cold February.

Table 53. REDSHANK: WINTER MAXIMA AT MAIN RESORTS

	88-89	89-90	90-91	91-92	92-93	(Mth)	Average
International							
Dee (Eng/Wales)	8,035	7,692	7,330	9,322	7,174	(Dec)	7,910
Morecambe Bay	7,151	6,763	6,379	5,756	7,272	(Nov)	6,664
Humber	(2,671)	(5,208)	(4,776)	4,219	(6,777)	(Jan)	5,245
Thames	(5,160)	6,040	(4,569)	(4,014)	2,567	(Mar)	4,584
Medway	(5,087)	(4,664)	3,450	5,355	3,073	(Mar)	4,325
Mersey	2,930	4,458	4,335	4,578	3,823	(Jan)	4,024
Forth	3,464	3,563	4,393	4,526	3,895	(Feb)	3,968
Wash	4,619	3,497	(3,872)	2,391	3,787	(Nov)	3,633
Inner Moray Fth	2,962	3,664	2,827	(2,654)	3,194	(Jan)	3,161
Solway	(1,851)	(1,966)	2,049	(3,127)	(3,192)	(Nov)	2,789
Severn	2,627	(1,614)	2,166	(2,841)	2,924	(Nov)	2,639
Lindisfarne	3,100	3,600	2,600	(1,045)	1,157	(Nov)	2,614
Strangford Lo.	2,809	2,771	2,420	2,345	2,336	(Dec)	2,536
Swale	(3,714)	(1,552)	(1,472)	1,817	1,463	(Dec)	2,331
Montrose Basin	1,983	2,530	2,717	2,202	2,049	(Mar)	2,296
Inner Clyde	2,243	1,546	2,441	1,817	1,998	(Jan)	2,009
Belfast Lo.	1,646	2,153	1,043	2,188	2,061	(Dec)	1,818
Cromarty Fth	1,829	1,168	2,304	1,634	1,339	(Feb)	1,654
Ribble	1,449	1,151	1,717	2,013	1,909	(Nov)	1,647
Alde complex	1,128	(1,458)	1,784	1,114	2,259	(Dec)	1,571
Deben	1,903	1,657	(1,191)	1,089	1,432	(Nov)	1,520
Great Britain							
Tay	1,051	2,339	711	1,152	2,236	(Mar)	1,497
Poole	1,997	858	1,012	1,300	2,178	(Jan)	1,469
Duddon	1,878	1,219	1,043	1,319	1,639	(Feb)	1,419
Orwell	1,373	1,243	1,574	1,531	862	(Dec)	1,316
Blackwater	598	(1,197)	1,023	1,728	(1,925)	(Jan)	1,294
Colne	1,247	1,288	1,152	1,332	1,391	(Mar)	1,282
Cleddau	1,603	1,629	1,326	729	911	(Dec)	1,239
Stour	905	1,185	1,478	1,279	1,331	(Mar)	1,235
Tees	1,319	(888)	1,264	986	960	(Dec)	1,132
Chichester Hbr	1,770	(1,595)	1,718	759	1,267	(Dec)	1,123
Northern Ireland							
Outer Ards*	847	1,267	863	737	779	(Jan)	898
Dundrum Bay	582	(725)	568	707	1,049	(Nov)	726
Lo. Foyle	800	597	730	(735)	634	(Feb)	699
Carlingford Lo.	938	810	557	388	688	(Feb)	676
South Down*	-	654	-	-	-		654
Larne Lo.	327	566	363	390	429	(Feb)	415

GREENSHANK
Tringa nebularia

International importance:		3,000
Great Britain importance:		+*
All-Ireland importance:		+*

GB maximum:	220	Dec	Trend	not available
NI maximum:	76	Jan		

During the 1992-93 winter, UK totals of Greenshank were unexceptional, with around 250 recorded in each month except March. Winter peak counts over 20 birds were made only at Strangford Lough (39 in November), Tamar complex (26 in December) and the Taw/Torridge (25 in January). Recorded UK totals of Greenshank during autumn passage normally greatly exceed those made during the winter. In autumn 1992, recorded numbers were on the low side in comparison to recent averages. Peak passage counts over 50 were made at the Wash (182 in July), Blackwater (136 in September), Medway (106 in August), Chichester Harbour (85 in September), Morecambe Bay (84 in September), Langstone Harbour (68 in August) and Strangford Lough (67 in October). Most of these high passage counts are usually recorded in autumn and at sites in SE England.

For only the second time, movement by Greenshank between the UK and south of the Sahara was confirmed. An adult ringed near Portsmouth in July 1985 was found freshly dead in Ghana in January 1992. (Mead *et al.* 1993).

GREEN SANDPIPER
Tringa ochropus

GB maximum:	98	Mar
NI maximum:	0	

UK totals of Green Sandpiper at coastal/estuarine sites were marginally higher than recent averages in February and March 1993. Similarly, at inland sites, UK totals in these two months were higher than in the corresponding months of the previous winter with the totals for inland sites slightly higher than those for coastal/estuarine sites. Sites holding five or more birds in the 1992-93 winter were the Thames (10 in February and March), Stour Valley# (seven in January), Colne (seven in February), Tamar complex (six in November and February), Swale (six in February) and Crouch/Roach (five in February). As usual, the main wintering areas are in S England. Passage period UK totals at coastal/estuarine sites were similar to recent averages with the autumn figures, as usual, exceeding those from the winter period. For the second consecutive year however no sites recorded counts over 50 birds, with the maximum being 25, recorded at the Thames and at the Swale.

Green Sandpipers leave their breeding grounds very early in the year, with birds on passage in Britain from late June to early October and peak numbers in August (Prater 1981). The British and Irish wintering population has been estimated at between 500 and 1000 individuals (Lack 1986) which suggests that only 10% of the population is recorded by WeBS counts, presumably because the preferred habitat includes small ponds and ditches. Ringing recoveries suggest that the majority of passage birds fly further south to Africa and that Britain's wintering populations comprise largely Scandinavian breeding birds (Mead & Clark 1993).

A recent study has found that Green Sandpipers have very high return rates from one winter to the next (83.5%), being very site faithful in mid-winter (Smith *et al.* 1992). Interestingly the return rates of Green Sandpipers are negatively correlated with the number of night frosts which occur in the first winter (Smith, Reed & Trevis 1992).

COMMON SANDPIPER
Actitis hypoleucos

GB maximum:	32	Nov
NI maximum:	0	

The wintering distribution of this species has a strong southerly bias which is illustrated in the 1992-93 counts. During the 1992-93 winter peak counts of greater than two birds were made at the Tamar complex (six in November and February), Medway (six in January), Barn Elms Resr.# (four in November), Long Ridge Reservoirs# (three in December) and Exe (three in November and March). UK totals for the winter were similar to recent averages. As usual the highest peak count made during 1992-93 was at Morecambe Bay (106 in July). In addition passage peak counts exceeding 50 were made at the Wash (62 in August) and Severn (51 in July).

The British and Irish wintering population has been estimated at around 100 individuals (Prater 1981; Lack 1986) and perhaps as a consequence very little work has been carried out on British wintering Common Sandpipers. Evidence from ringing recoveries (Mead & Clark 1993) suggests that the British breeding population winters in Africa, with most authors (Prater 1981; Lack 1986) suggesting an area south of the Sahara. The furthest British recovery was a bird ringed in Essex and subsequently shot in Guinea-Bissau. The British wintering population is made up of Scandinavian waders with 23 ringed birds confirming British-Scandinavian movements. The distribution maps from the two British atlases (winter and breeding) (Lack 1986, Gibbons et al. 1993) show that breeding and wintering birds occupy totally different geographical areas. Breeding birds prefer upland areas of Wales, N. England and Scotland, whereas wintering birds occur in lowland England and Wales, predominantly in coastal areas.

TURNSTONE
Arenaria interpres

International importance:		700
Great Britain importance:		650
All-Ireland importance:		225

GB maximum: 15,934 Jan
NI maximum: 2,223 Jan

Trend	88-89	89-90	90-91	91-92	92-93
UK	144	152	141	153	134

In 1992-93 the UK winter index dropped by 12% to reach its lowest value for eight years. The majority of the wintering population in the UK, however, is not covered by WeBS counts since around 80% occur on non-estuarine coasts, many of which are not included in the survey. All internationally important sites which are included within the current counts plus those of national importance and those important in an all-Ireland context are shown in Table 54. Other sites of importance may remain uncounted. At only two of these major sites (Morecambe Bay and Belfast Lough) was the winter 1992-93 peak above the recent average. The greatest decline in numbers was noted at the Wash.

During passage periods in 1992-93, peak counts exceeding 700 birds were noted at seven sites. Sites with a peak in autumn were Morecambe Bay (1,645 in October), Thames (1,551 in September), Thanet* (1,244 in August), Medway (1,176 in September) and Wash (739 in August). This list demonstrates the importance of the N Kent and S Essex region for migrant Turnstones. Peak passage counts were recorded in the spring at Guernsey* (812 in April) and the Forth (701 in April). For the first time in three years the peak count at Morecambe Bay was in the autumn rather than the spring.

Table 54. TURNSTONE: WINTER MAXIMA AT MAIN RESORTS

	88-89	89-90	90-91	91-92	92-93	(Mth)	Average
International							
Morecambe Bay	1,647	1,651	1,944	1,721	2,086	(Jan)	1,809
Outer Ards*	1,775	2,336	1,612	(1,207)	1,163	(Jan)	1,721
Thanet*	1,284	1,144	1,253	1,342	949	(Nov)	1,194
South Down*	-	1,190	-	-	-		1,190
Forth	1,184	(869)	(1,188)	1,082	1,066	(Jan)	1,130
Thames	(681)	(595)	766	1,379	(665)	(Nov)	1,072
Wash	1,282	967	(1,131)	896	457	(Mar)	946
Dee (Eng/Wales)	960	1,185	853	780	765	(Nov)	908
Guernsey*	602	(664)	936	615	(565)	(Mar)	717
Northern Ireland							
Belfast Lo.	575	778	877	476	770	(Jan)	695
Strangford Lo.	437	455	326	346	312	(Nov)	375
Tyrella/Minerstown*	-	449	186	233	-		289

KINGFISHER
Alcedo atthis

International importance:		?
Great Britain importance:		?
All-Ireland importance:		?

GB maximum: 271 Sep
NI maximum: 1 Sep/Nov

Trend: not available

The Kingfisher's quiet habits and preference for slow flowing rivers (Taylor 1986) mean that, despite its dazzling appearance, it is recorded only in small numbers by WeBS. The 1992-93 peak count is considerably higher than the 1991-92 figure, and March numbers were also well up, possibly due to the relatively mild winter. Harsh winter conditions are known to cause severe mortality amongst Kingfisher populations, though it is interesting to note that the breeding distribution is gradually spreading further north (Mead 1993). Estimated breeding numbers of between 3,300 and 5,500 pairs are present in Britain, and hence apparent trends or changes noted by the winter counts are subject to a great deal of uncertainty since only a small proportion of the population is ever recorded. The following sites held 5 or more birds during 1992-93: Tamar Estuary (10, November, *cf.* 0), Poole Harbour (10, October, *cf.* 3), Somerset Levels (8, October, *cf.* 5), Wraysbury Gravel Pits (7, October, *cf.* 8), River Ouse: Southease to Lewes (6, October, not counted in 1991-92), Stour Estuary (6, September, *cf.* 0), Stockers Lake (6, March, *cf.* 1), Cheshunt Gravel Pit (6, March, *cf.* 5), Theale Gravel Pits (6, February, *cf.* 2), Drakelow Gravel Pit (6, October, *cf.* 0), Cotswold Water Park West (5, March, *cf.* 2), Tring Reservoirs (5 November, *cf.* 2), Frisby Gravel Pits (5, September, *cf.* 3), Gunthorpe Gravel Pits (5, September, *cf.* 2), Lancaster Canal (5, January, *cf.* 0) and Eglwys Nunydd Reservoir (5, October, *cf.* 3).

ADDITIONAL SPECIES

Several species of waterfowl, although they occur regularly in Britain, are only recorded in very small numbers by WeBS. These include scarce visitors and rarities (see e.g. Rogers and the Rarities Committee 1993) and escaped or feral birds. In several cases, it is not possible to identify the origin of individual birds with confidence, and, following the recommendations of Vinicombe *et al.* (1993), all WeBS records are presented here for completeness.

LESSER WHITE-FRONTED GOOSE *Anser erythropus*

Birds were recorded at three sites in 1992-93: one was seen on Alton Water in February and April, another on the Duddon in August, September and November in 1992, although the bird seen on Derwent Water in February and March 1993 may have been the same bird as that on the Duddon.

RED-BREASTED GOOSE *Branta ruficollis*

Singles were noted at Fisherwick and Elford Gravel Pits in September and October, and on the Blackwater Estuary in February.

AMERICAN WIGEON *Anas americana*

Possibly as many as seven birds were found at six sites in 1992-93, although some records probably relate to the same individuals. Singles were recorded at Loch of Skene, Stenhouse Reservoir and Loch of Loriston with two on the Firth of Fourth, all during mid-winter, while there were MArch records of single birds at Martin Mere and Foryd Bay.

RING-NECKED DUCK *Aythya collaris*

Numbers of this species have increased in recent years and birds were recorded widely throughout the UK, with some individuals having returned to the same site for several winters in succession. Singles were found at Upper Lough Erne, Loe Pool, Drift Reservoir, Timsbury Lake, Broadlands Estate, Fairburn Ings, Winfields Pond and Whinfell Tarn, while two birds found on Loch Morlich in November moved to Loch Vaa in December.

FERRUGINOUS DUCK *Aythya nyroca*

A single bird was recorded at Wicken Fen Gravel Pit in November while another overwintered at Kingsbury and Coton Pools.

KING EIDER *Somateria spectabilis*

A lone bird was recorded amongst the large numbers of sea-duck at Loch Fleet in October, and two were recorded by RSPB/BP surveys on the Moray Firth.

SURF SCOTER *Melanitta perspicillata*

Four birds were discerned amongst their European cousins in the Firth of Forth in January, with another two on the Eden Estuary in the same month. At least a further three were recorded by RSPB/BP surveys in the Moray Firth (Evans 1993).

WOOD SANDPIPER *Tringa glareola*

Recorded at 11 sites with the maximum count of two birds noted at the Tamar complex in May and at the North Norfolk Marshes in September. No birds were recorded in the 1992-93 winter period.

KENTISH PLOVER *Charadrius alexandrinus*

The winter record of an individual at Morecambe Bay in December, January and February is highly unseasonal. One bird on the Exe in April was at a more typical time of year.

LITTLE RINGED PLOVER *Charadrius dubius*

Recorded at 32 sites, many inland, with the maximum counts being six birds at North Norfolk Marshes in August and the same number at Sennowe Park, Lake Guist# in July and August.

GREY PHALAROPE *Phalaropus fulicarius*

Recorded only at Rosehearty-Fraserburgh* (in January), Morecambe Bay, Llangorse Lake# and Loch an Tiumpan (Lewis)#, all in September. Single individuals only.

PECTORAL SANDPIPER *Calidris melanotos*

All counts were made in September with two on the Tees and one bird recorded at both North Norfolk Marshes and the Avon estuary.

WOODCOCK *Scolopax rusticola*

Only singletons were recorded from nine sites between October and March.

RED-NECKED PHALAROPE *Phalaropus lobatus*

One at the Severn Estuary in September.

TEMMINCK'S STINT *Calidris temminckii*

One at the Tees in July.

BUFF-BREASTED SANDPIPER *Tryngites subruficollis*

One at Netherfield Gravel Pits# in September.

LESSER YELLOWLEGS *Tringa flavipes*

The October count at Loch Ordais (Lewis)# included one individual.

PRINCIPAL SITES

Table 55 lists the principal sites in terms of overall waterfowl numbers in the UK as recorded by WeBS, including all internationally important sites. All sites regularly holding a total of at least 10,000 waterfowl (i.e. divers, grebes, Cormorant, Grey Heron, wildfowl, waders and rails), or, where wader data for the last five seasons are not available (i.e. for inland sites), 10,000 or more wildfowl (i.e. divers, grebes, Cormorant, wildfowl and Coot) are included here. These are ranked according to their average maxima over the five-year period 1988-89 to 1992-93. Sites supporting one or more species with a population level of international importance (see Appendix 1), according to average maxima calculated over the five year period 1988-89 to 1992-93, are also included.

It is important to note that the ranking of sites given in Table 55 relates to waterfowl numbers, rather than conservation importance (see *Interpretation of Waterfowl Counts*). Also, some sites which may be of critical importance to certain waterfowl species or populations will not be included in this list, for example, sites that are important only in times of severe weather or during migratory periods. The locations of the sites in Table 55 are given in Appendix 2.

The peak counts at each site are calculated by summing the highest count for each individual species during the winter season, irrespective of the month in which it occurred. The table shows the average peak counts at each site over the period 1988-89 to 1992-93, and the peak counts of all waterfowl, wildfowl and waders in 1992-93 in successive columns. For most inland sites, the numbers of waders present has only been recorded for the past two years. In these cases, the five year average value is calculated using wildfowl data alone, and is marked in the table by an asterisk. The number of Internationally Important Populations (IIP) and corresponding species codes are given in the final two columns.

Note that data for the "wildfowl" species first recorded in WeBS databases in 1991-92 have been included in calculating site totals of waterfowl and wildfowl in Table 55 for 1992-93 only. While totals for previous seasons would have been higher if these species were included for those years also, the numbers involved are, in many cases, negligible. The maximum increase for any site as a result of including these extra species in 1992-93 was 375 birds. The average increase per site was just 10 birds, although 78% of sites recorded increases of less than this. Only data collected during WeBS monthly counts and the censuses of Pink-footed and Greylag Geese are included in calculating site totals. Additional counts, such as those of sea-ducks on the Moray Firth, made using different methodologies, are not currently incorporated into the WeBS databases.

It is also worth noting that, since the production of the last report (Cranswick *et al.* 1992), revised qualifying levels for international importance have been published for some species (Rose & Scott 1993), e.g. Gadwall 1% international level has increased from 120 to 250. Hence, some sites may appear to have declined in importance for particular species when compared to previous years.

Though the table requires careful interpretation, it does serve to identify many of the UK's important wetlands, and some of the species for which these sites have special value. Readers should refer to the sections on *Interpretation of Waterfowl Counts* and *Data Presentation* for guidance.

Wildfowl numbers inevitably fluctuate from year to year, so detecting significant increases or decreases is difficult. However, of sites that regularly support in excess of 20,000 waterfowl, the following held peak counts that were at least 30% above or below their respective five year averages in 1992-93: the Blackwater Estuary (+68%), Breydon Water (+43%), the Stour Estuary (30%) and the Alt Estuary (-71%). In addition, the increases on the Dee Estuary (England/Wales) and the Wash were also remarkable in terms of the absolute numbers of birds, although representing only 27% and 18% increases respectively.

Variations in just total wildfowl or total wader numbers may be masked when considering all waterfowl combined. Of sites regularly supporting in excess of 10,000 wildfowl, counts at the following sites were 30% above or below their respective five year averages: Cameron Reservoir (+54%), Montrose Basin (+31%), Tay Estuary (-77%), Slains Loch (-76%), Lough Foyle (-47%), Dupplin Lochs (-35%), Lindisfarne (-35%) and Abberton Reservoir (-31%). Similarly, for sites which regularly hold 10,000 or more waders, the following recorded peak counts of 30% above or below their respective five year averages: Langstone Harbour (+92%), Inner Moray Firth (+51%), Breydon Water (+44%), Alde Complex (43%), Carmarthen Bay (+40%), Chichester Harbour (+37%), Dee Estuary (+37%), Poole Harbour (+37%) and Medway Estuary (-74%).

The peak 1992-93 count of wildfowl at many of the key sites in Table 55 was either on a par with or, for some sites, much lower than, their respective five year average. As a result of the poor breeding success of many Arctic-nesting geese, the populations were, for some species, lower than in recent years. There were no particularly large concentrations of Pink-footed Geese in 1992-93 and the count at Dupplin Lochs was less than half that of 1991-92. Numbers of Dark-bellied Brent Geese were also lower, which, combined with fewer Pinkfeet, contributed to the low count on the Wash. Although the 1992-93 peak of wildfowl of the Ribble Estuary was very similar to the site's five year average, it was almost 50,000 less than the record count of 1991-92, largely due to the "lesser" numbers of Wigeon. Numbers on the Mersey and the Dee (England/Wales) Estuaries were also lower than in recent seasons, mostly as a result of fewer Teal and Pintail at these sites. Smaller numbers of all four species mentioned thus far contributed to the 1992-93 total for the North Norfolk Marshes being the lowest for over five years and less than half the figure recorded in 1991-92. A similar pattern of total numbers was observed at Lough Foyle, as a result of much fewer Wigeon and Light-bellied Brent Geese at the site in 1992-93. Some of the most notable declines at inland sites were at Loughs Neagh and Beg, which held far fewer diving ducks than in recent years, and Abberton Reservoir, where numbers of several species were lower than normal. Montrose Basin and the Swale Estuary were some of the few sites to record more wildfowl than in recent years. Most notable, however, were the increases on the Somerset Levels, which must be rich reward for the successful management of this area, especially West Sedgemoor, in recent years. Numbers of Teal and Wigeon here have risen dramatically in recent years and the total of over 80,000 waterfowl in 1992-93 indicates that this area will almost certainly feature much higher in Table 55 in years to come.

Most sites with peak counts averaging over 10,000 waders over the past five winters recorded above average peaks in the 1992-93 winter. High numbers of Lapwing and Golden Plover were present at many sites, giving rise to the record high UK totals recorded. At the Blackwater, high counts of Lapwing, Golden Plover and Dunlin increased the "all species" total to more than double the average peak of the previous five winters. High numbers of Lapwing were also recorded on the adjacent estuaries of Hamford Water and the Stour. The overall numbers of waders recorded at each of these three estuaries in eastern England has increased every winter for the past three years. Also, for the third winter in succession, Golden Plover and Lapwing were present in above average numbers at Breydon Water, giving rise to a further increase in the all waders total. Much further west at Carmarthen Bay, exceptionally large numbers of Golden Plover were recorded, more than double the previous average all wader total. Of those major sites whose winter peaks average more than 100,000 waders in recent years, the Dee (England/Wales) registered the greatest change in total numbers. The peak count of waders in 1992-93 was more than 50% higher than the average of the previous five winters, due mainly to increased numbers of Oystercatcher and Knot. At the nearby Alt, overall numbers of waders declined more in 1992-93 than at any other major site. Dunlin and especially Knot numbers were particularly low on the Alt. Movement of Knot between the Alt and the Dee is well known and probably accounts for most of these observed results. Montrose Basin was the only other major site recording overall wader totals well below those of recent years. The 40% drop in the winter peak count at this estuary in eastern Scotland was due largely to unusually low numbers of Dunlin and Golden Plover.

The general trend at inland wetlands was for the 1992-93 winter peak of waders to exceed that of the previous winter with Lapwing, in particular, and Golden Plover recorded in larger numbers. A notable exception to this pattern was Loughs Neagh and Beg where the 1992-93 peak was less than half that of the previous winter due to much lower numbers of Lapwing and Golden Plover. It will be interesting to see how counts in future winters will enable us to draw firm conclusions about the importance of sites for inland wader populations.

SPECIES CODES

Species	Code	Species	Code
Little Grebe	LG	Scaup	SP
Great Crested Grebe	GG	Eider	E
Cormorant	CA	Long-tailed Duck	LN
Mute Swan	MS	Goldeneye	GN
Bewick's Swan	BS	Red-breasted Merganser	RM
Whooper Swan	WS	Goosander	GD
Pink-footed Goose	PG	Coot	CO
European White-fronted Goose	EW	Oystercatcher	OC
Greenland White-fronted Goose	NW	Avocet	AV
Greylag Goose	GJ	Little Ringed Plover	LP
Canada Goose	CG	Ringed Plover	RP
Barnacle Goose	BY	Golden Plover	GP
Dark-bellied Brent Goose	DB	Grey Plover	GV
Light-bellied Brent Goose	PB	Lapwing	L
Shelduck	SU	Knot	KN
Wigeon	WN	Sanderling	SS
Gadwall	GA	Dunlin	DN
Teal	T	Black-tailed Godwit	BW
Mallard	MA	Bar-tailed Godwit	BA
Pintail	PT	Whimbrel	WM
Shoveler	SV	Curlew	CU
Pochard	PO	Redshank	RK
Tufted Duck	TU	Turnstone	TT

NB *Not every species covered by WeBS has a corresponding qualifying level for international importance (see Appendix 1). Hence these species do not feature in this table.*

TABLE 55. PRINCIPAL WATERFOWL SITES IN THE UK, 1988-89 TO 1992-93

Site name	5 Yr Mean Waterfowl	1992-93 Waterfowl	1992-93 Wildfowl	1992-93 Waders	†IIP	Species codes
Wash	343,866	407,748	54,731	353,017	13	PG,DB,SU,PT,OC,GV,L,KN,DN,BA,CU,RK,TT
Ribble Est.	237,721	230,741	85,838	144,903	15	BS,WS,SU,WN,T,PT,OC,GV,L,KN,SS,DN,BW,BA,RK
Morecambe Bay	223,068	219,701	32,317	187,384	12	PG,SU,PT,OC,GV,L,KN,DN,BA,CU,RK,TT
Dee (Eng/Wales)	145,760	186,422	28,124	158,298	12	SU,T,PT,OC,GV,KN,DN,BW,BA,CU,RK,TT
Humber Est.	143,065	142,997	21,008	121,989	9	DB,SU,GP,GV,L,KN,DN,BA,RK
Thames Est.	142,886	151,533	28,164	123,369	10	DB,SU,OC,RP,GV,KN,DN,BA,RK,TT
Solway Est.	123,935	123,572	39,799	83,773	9	WS,PG,BY,PT,SP,OC,KN,BA,CU,RK
Lo. Neagh/Beg	*97,658	95,713	84,755	10,958	6	WS,BS,PO,TU,SP,GN
Mersey Est.	82,942	80,090	30,571	49,519	5	SU,T,PT,DN,RK
Severn Est.	80,941	69,538	18,119	51,419	5	BS,SU,GA,DN,RK
Forth Est.	76,914	74,237	35,001	39,236	6	PG,SU,KN,BA,RK,TT
Medway Est.	72,290	70,783	16,735	54,048	7	DB,SU,PT,RP,GV,DN,RK
N. Norfolk Marshes	70,351	62,469	34,224	28,245	6	PG,DB,WN,PT,KN,BA
Strangford Lo.	61,081	53,778	18,618	35,160	3	BI,KN,RK
Blackwater Est.	60,318	101,152	23,179	77,973	5	DB,SU,GV,DN,BW
Swale Est.	59,848	71,459	26,308	45,151	6	DB,WN,GV,KN,BW,RK
Ouse Washes	*52,654	72,465	57,404	15,061	7	BS,WS,WN,GA,T,PT,SV
Chichester Hbr	52,453	54,619	16,732	37,887	5	DB,RP,GV,DN,BA
Langstone Hbr	52,236	47,418	10,215	37,203	2	DB,DN
Lindisfarne	50,710	45,283	14,873	30,410	5	GJ,WN,RP,BA,RK
Inner Moray Fth	49,308	50,926	22,144	28,782	6	PG,GJ,WN,RM,BA,RK
Montrose Basin	47,078	52,761	42,864	9,897	2	PG,RK
Alt Est.	45,793	13,232	1,678	11,554	2	KN,BA
Stour Est.	43,908	57,143	10,810	46,333	3	GV,DN,BW
Lo. of Strathbeg	*39,812	37,691	37,691	-	3	WS,PG,GJ
Dupplin Lo.	*39,238	25,560	25,500	60	1	PG
Lo. Foyle	38,878	32,071	12,187	19,884	3	WS,WN,BA
Burry Inlet	37,614	27,777	5,497	22,280	2	PT,OC
Colne Est.	35,336	43,327	13,227	30,100	1	DB
Duddon Est.	34,929	43,068	8,016	35,052	2	PT,KN
Abberton Rsr	*34,772	24,072	24,072	-	3	GA,T,SV
Lo. Leven	*32,279	33,188	32,655	533	2	PG,SV
West Water Rsr	*32,200	25,329	25,269	60	1	PG
Hamford Water	31,317	38,018	9,987	28,031	4	DB,RP,GV,BW
Dengie	29,843	33,760	5,114	28,646	2	GV,KN
Dornoch Fth	29,243	31,577	21,072	10,505	2	GJ,WN,
Cromarty Fth	27,518	22,544	13,452	9,092	6	WS,PG,GJ,WN,BA,RK
Tay Est.	26,824	16,456	3,564	12,892	3	E,BA,RK
Crouch/Roach Est.	26,767	28,210	8,953	19,257	1	DB
Martin Mere	*26,229	29,401	27,964	1,437	4	BS,WS,WN,PT
Exe Est.	23,150	21,573	7,038	14,535		
Poole Hbr	22,821	28,437	8,843	19,594	2	SU,BW
Alde Complex	22,679	27,105	8,857	18,248	2	AV,RK
Lo. Eye	*22,642	20,052	20,044	8	3	WS,PG,GJ
Inner Clyde	22,252	24,040	5,936	18,104	1	RK
Breydon Water	21,428	30,731	7,037	23,694	1	BS
Orwell Est.	21,018	22,461	7,002	15,459		
Rutland Water	*20,959	19,551	16,222	3,329	2	GA,SV
Tees Est.	20,522	19,511	5,723	13,788	1	KN
Dinnet Lo.	*19,957	23,533	23,533	-	1	GJ
Southampton Water	19,751	19,966	8,100	11,866		
Belfast Lo.	19,459	22,863	5,366	17,497	1	RK
Carmarthen Bay	19,083	26,058	933	25,125		
Outer Ards	18,583	16,822	511	16,311	2	RP,TT
Slains Lo.	*18,161	4,360	4,360	-	1	PG
Lo. of Skene	*18,156	21,020	21,020	-	2	WS,GJ

Site name	5 Yr Mean Waterfowl	1992-93 Waterfowl	1992-93 Wildfowl	1992-93 Waders	†IIP	Species codes
Fleet/Wey	17,928	18,761	16,756	2,005		
Cleddau Est.	17,254	15,769	7,314	8,455		
Hule Moss	*17,161	17,373	17,354	19	1	PG
Wigtown Bay	16,603	11,870	5,292	6,578	1	PG
Lower Derwent Ings	*16,398	37,921	18,344	19,577		
Deben Est.	15,336	14,996	4,873	10,123	1	RK
Eden Est.	15,066	15,745	4,359	11,386		
Carsebreck/Rhynd Lo.	*14,369	13,289	11,300	1,989	1	PG
Pagham Hbr	13,646	14,919	7,298	7,621	1	DB
Portsmouth Hbr	13,611	10,600	4,104	6,496	1	DB
NW Solent	13,140	13,997	5,297	8,700	1	DB
Lo. of Harray	*12,701	12,428	9,651	2,777	2	WS,GJ
Tamar Complex	11,978	11,940	2,909	9,031		
Lavan Sands	11,769	10,297	2,314	7,983		
Taw/Torridge Est.	11,131	10,994	2,969	8,025		
Cameron Rsr	*10,726	17,015	16,581	434	1	PG
Castle Lo., L'maben	*12,558	(3,000)	(3,000)	-	1	PG
Dundrum Bay	10,403	10,297	2,048	8,249		
Dyfi Est.	10,232	9,780	6,046	3,734		
Ythan Est.	9,971	10,663	2,855	7,808		
Chew Valley Lake	*9,791	23,036	8,934	14,102	2	GA,SV
Fala Flow	*9,220	4,881	4,845	36	1	PG
Lo. Fleet Complex	9,169	7,503	3,862	3,641		
Somerset Levels	*9,031	80,370	24,357	56,013	1	T
Irvine Est.	8,845	10,048	3,862	6,186		
South Down	8,809	8,809	-	8,809	2	RP,TT
Nene Washes	*8,799	25,278	12,903	12,375	1	BS
Blyth Est. (Suffolk)	8,350	12,117	2,906	9,211		
Rye Hbr/Pett Levels	8,209	7,654	2,425	5,229		
Camel Est.	8,142	10,729	349	10,380		
Lo. Spynie	*8,132	8,140	8,139	1	1	GJ
Thanet Coast	8,087	8,087	2,036	6,051	1	TT
Beaulieu Est.	7,821	9,208	3,575	5,633		
Newtown Est.	7,594	7,390	4,263	3,127		
Tynighame Est.	6,985	8,104	2,261	5,843		
Avon Valley (Mid)	*6,936	9,287	9,287	-	2	BS,GA
Carlingford Lo.	6,767	6,081	2,263	3,818		
Lo. of Kinnordy	*6,227	5,709	5,563	146	1	PG
Upper Lo. Erne	*6,074	5,492	4,047	1,445	1	WS
Drummond Pond	*6,009	11,053	9,071	1,982	1	GJ
Haddo House Lo.	*5,965	5,702	5,696	6	1	GJ
Lo. Indaal	5,921	5,428	3,415	2,013		
Inland Sea	5,812	7,187	2,018	5,169		
Lo. of Lintrathen	*5,756	6,346	6,232	114	1	GJ
Pegwell Bay	5,661	6,848	1,248	5,600		
Swansea Bay	5,204	3,950	124	(3,826)		
Lo. Larne	5,056	4,036	1,963	2,073		
Hayle Est.	4,895	8,529	1,248	7,281		
Fal Complex	4,832	5,924	971	4,953		
Lo. Ryan	4,665	4,936	2,245	2,691		
Gladhouse Rsr	*4,470	3,676	3,673	3	1	PG
Lo. Tullybelton	*4,370	5,800	5,800	-	1	PG
Christchurch Hbr	4,282	6,006	771	5,235		
Auchencairn Bay	4,208	2,825	590	2,235		
Traeth Bach	4,205	2,699	2,396	1,303		
Irt/Mite/Esk Est.	4,073	4,311	1,722	2,589		
Crombie Lo.	*3,972	3,953	3,917	36	1	PG
Clwyd Est.	3,956	4,718	884	3,834		

Site name	5 Yr Mean Waterfowl	1992-93 Waterfowl	1992-93 Wildfowl	1992-93 Waders	⁺IIP	Species codes
Foryd Bay	3,901	4,304	2,744	1,560		
Lo. Mahaick	*3,772	800	800	-	I	PG
Kingsbridge Est.	3,746	4,119	2,117	2,002		
Conwy Est.	3,717	3,297	1,074	3,223		
Bann Est.	3,696	3,872	662	3,210		
Cowgill Rsr	*3,579	6,700	6,700	-	I	PG
Cefni Est.	3,305	3,041	1,147	1,894		
Hoselaw Lo.	*3,274	907	905	2	I	GJ
Lo. Lomond	*3,137	3,434	3,434	-	I	NW
Guernsey Coast	3,123	2,993	169	2,824	I	TT
Holburn Moss	*3,099	3,404	3,404	-	I	GJ
Lake of Menteith	*3,063	80	80	-	I	PG
Lo. Ken	*2,910	(323)	(323)	-	I	NW
Lo. Gruinart	2,867	3,086	1,043	2,041		
Brading Hbr	2,815	4,117	1,202	2,915		
Adur Est.	2,810	4,153	63	4,090		
Red Wharf Bay	2,704	2,245	589	1,656		
Fedderate Rsr	*2,605	-	-	-	I	GJ
Lour	*2,605	-	-	-	I	PG
Luce Bay	2,594	3,329	554	2,775		
Thorpe WP	*2,548	2,091	2,091	-	I	GA
Lo. Clunie	*2,547	2,718	2,718	-	I	GJ
Berney Marshes	*2,532	8,407	4,537	3,870	I	BS
Braint Est.	2,519	1,378	256	1,122		
Ballo Rsr	*2,316	3,212	3,212	-	I	GJ
Cuckmere Est.	2,282	2,217	1,592	625		
Gadloch	*2,247	2,958	2,958	-	I	GJ
Corby Lo.	*2,099	1,530	1,273	257	I	GJ
Mawddach Est.	2,032	1,612	873	739		
Yar Est.	2,012	1,850	1,481	369		
Tweed Est.	1,987	2,287	1,251	1,036		
Hunterston Est.	1,930	1,643	1,176	467		
Lower Bogrotten	*1,883	3,000	3,000	-	I	GJ
Dysynni Est.	1,855	2,737	1,842	895		
Rough Firth	1,795	2,498	419	2,079		
Gartmorn Dam	*1,719	1,200	1,200	-	I	GJ
Newhaven Est.	1,550	1,953	25	1,928		
St Benet's Levels	*1,499	3,368	898	2,470	I	BS
Gunton Park Lakes	*1,494	1,154	1,154	0	I	GA
Lo. of the Lowes	*1,452	3,040	3,040	-	I	GJ
Ogmore Est.	1,434	1,434	560	874		
Blyth Est. (N'berland)	1,395	1,177	226	951		
Kircudbright Bay	1,363	1,434	739	695		
Medina Est.	1,353	2,049	419	1,630		
Coquet	1,341	1,600	405	1,195		
Teifi Est.	1,320	1,215	634	581		
Lossie Est.	1,314	1,681	860	821		
Axe Est.	1,279	2,480	311	2,169		
Lo. Garten	*1,169	1,057	1,057	-	I	GJ
Otter Est.	1,076	1,120	1,028	92		
Machrihanish	*1,057	1,110	1,110	-	I	NW
Dee Est. (Scotland)	1,031	2,052	881	1,171		
Nyfer Est.	1,029	548	194	354		
Rhunahaorine	*1,010	726	726	-	I	NW
Lo. Gilp	1,003	795	290	505		
Dulas Bay	948	782	81	701		
Plym Est.	904	685	106	579		
Avon Est.	776	567	357	210		

Site name	5 Yr Mean Waterfowl	1992-93 Waterfowl	1992-93 Wildfowl	1992-93 Waders	†IIP	Species codes
Deveron Est.	690	656	174	482		
Erme Est.	677	608	481	127		
Artro Est.	648	805	356	349		
Yealm Est.	647	783	560	223		
Wootton Est.	639	735	298	437		
Gannel Est.	624	569	65	504		
Don Est.	606	792	263	529		
Fleet Bay	604	602	181	421		
Linne Mhurich/Lo. na Cille	*542	542	542	-	1	NW
Teign Est.	425	292	131	161		
South Alnmouth	334	271	103	168		
Spey Est.	329	839	795	44		
Tyne Est.	301	-	-	-		
Dart Est.	247	96	96	-		
Afan Est.	239	239	83	156		
Fowey Est.	231	298	113	185		
Helford Est.	214	262	108	154		
Looe Est.	169	-	-	-		
Caithness Lo.	-	-	-	-	1	GJ
Islay	-	-	-	-	2	NW,BY
Walland Marsh	-	-	-	-	1	BS
SW Lancashire	-	-	-	-	1	PG
Coll	-	-	-	-	1	NW
Tiree	-	-	-	-	1	NW
Tay/Isla Valley	-	-	-	-	1	GJ
Stranraer Lo.	-	-	-	-	2	NW,GJ
Orkney	-	-	-	-	1	GJ
Bute Lochs	-	-	-	-	1	GJ

- *Indicates that no total count is available*

REFERENCES

Amat, J.A. & Soriguer, R.C. 1984. Kleptoparasitism of Coots by Gadwall. *Ornis Scand.* 15: 188-194.

Anon. 1993. *Biodiversity Challenge: an agenda for conservation action in the UK.* Butterfly Conservation, Friends of the Earth, Plantlife, The Royal Society for Nature Conservation - the Wildlife Trusts' partnership, Royal Society for the Protection of Birds and World Wide Fund for Nature.

Anon. 1994. *Biodiversity: the UK Action Plan.* Cm, 2428, HMSO, London.

Baatsen, R.G. 1990. Red-crested Pochard *Netta rufina* in the Cotswold Water Park. In: *Hobby 1990.* Wiltshire Ornithological Society, Salisbury.

Berg, A. 1992. Habitat selection by breeding Curlews *Numenius arquata* on mosaic farmland. *Ibis* 134: 355-360.

Berg, A. 1993. Food resources and foraging success of Curlews *Numenius arquata* in different farmland habitats. *Ornis Fenn.* 70: 22-31.

Bowler, J.M., Butler, L. & Rees, E.C. 1993. Bewick's and Whooper Swans *Cygnus columbianus bewickii* and *C. cygnus*: the 1992-93 season. *Wildfowl* 44: 191-198.

Brown, A.F. 1993. The status of the Golden Plover *Pluvialis apricaria* in the south Pennines. *Bird Study* 40: 196-202.

Bustnes, J.O. 1993. Exploitation of others' vigilance by the Common Eider *Somateria mollissima*. *Wildfowl* 44: 108-110.

Cadbury, J. & Kirby, J. 1993. Black-tailed Godwit. In: Gibbons, D.W., Reid, J.B. & Chapman, R. (Eds.) *The New Atlas of Breeding Birds in Britain and Ireland: 1988-1991.* Poyser, London.

Choudhury, S. & Owen, M. 1993. Migratory geese wintering on Islay. Assessing the Impact. Wetlands Advisory Service Report to The Scottish Office.

Chylarecki, P. & Kania, W. 1992a. Polygyny and polyandry in the mating system of the Little Stint *Calidris minuta*. *Wader Study Group Bull.* 64: 12.

Chylarecki, P. & Kania, W. 1992b. Research on waders at the Pysaina Mouth, western Taymyr in 1991. *Wader Study Group Bull.* 64: 11-12.

Clark, J.A., Clark, N.A., Gill, J.A. & Sutherland, W.J. 1993. *Wash Wader Ringing Group Report 1991-92.*

Clark, N.A. 1993. Wash Oystercatchers starving. *BTO News* 185: 1, 24.

Cox, S. 1986. Red-crested Pochard. In: Lack, P. (Ed.) *The Atlas of Wintering Birds in Britain and Ireland.* Poyser, Calton.

Cranswick, P.A. 1993a. An assessment of breeding success in the Dark-bellied Brent Goose *Branta bernicla bernicla* in 1992. WWT report to JNCC, Slimbridge, 2pp.

Cranswick, P.A. 1993b. Numbers of Dark-bellied Brent Geese in Britain, January/February 1993. WWT report to JNCC, Slimbridge, 4pp.

Cranswick, P.A., Kirby, J.S. & Waters, R.J. 1992. *Wildfowl and Wader Counts 1991-92.* WWT, Slimbridge.

Cresswell, W. 1993. Escape responses by Redshanks, *Tringa totanus*, on attack by avian predators. *Animal Behaviour* 46: 609-611.

Danielsen, F., Skov, H. & Durnick, J. 1993. Estimates of the wintering population of Red-throated Diver *Gavia stellata* and Black-throated Diver *Gavia arctica* in northwest Europe. Proc. *7th Nordic Congress of Ornithology*, 1990. Pp. 18-24.

Davidson, N.C., Laffoley, D.d'A., Doody, J.P., Way, L.S., Gordon, J., Key, R., Drake, C.M., Pienkowski, M.W., Mitchell, R. & Duff, K.L. 1991. *Nature Conservation and Estuaries in Great Britain.* Nature Conservancy Council, Peterborough.

Dekinga, A. & Piersma, T. 1993. Reconstructing diet composition on the basis of faeces in a mollusc-eating wader, the Knot *Calidris canutus*. *Bird Study* 40: 144-156.

Delany, S.N. 1992. Survey of introduced geese in Britain, summer 1991: provisional results. WWT report to JNCC, MAFF and the National Trust, Slimbridge, 38pp.

Delany, S.N. 1993a. Introduced and escaped geese in Britain in summer 1991. *Brit. Birds* 86: 591-599.

Delany, S.N. 1993b. Winter counts of waterfowl in the Cotswold Water Park 1989/90-1992/93. WWT report to JNCC, Slimbridge, 48pp.

Delany, S.N. 1993c. Red-crested Pochard. In: Gibbons, D.W., Reid, J.B. & Chapman, R. (Eds.) *The New Atlas of Breeding Birds in Britain and Ireland: 1988-1991.* Poyser, London.

Delany, S.N. in press. The 1991 survey of introduced geese. In: Andrews, J. (Ed.) *Britain's Birds in 1991-92: The Conservation and Monitoring Review.*

Delany, S.N. & Greenwood, J.J.D. 1993. The 1990 National Mute Swan Survey: provisional results. In: Andrews, J. & Carter, S.P. (Eds.) *Britain's Birds in 1990-91: The Conservation and Monitoring Review.* BTO/JNCC, Thetford.

Delany, S.N. and Ogilvie, M.A. 1993. Mute Swan. In: Gibbons, D.W., Reid, J.B. & Chapman, R. (Eds.) *The New Atlas of Breeding Birds in Britain and Ireland: 1988-1991.* Poyser, London.

Duncan, K. & Marquiss, M. 1993. The sex/age ratio, diving behaviour and habitat use of Goldeneye *Bucephela clangula* wintering in northeast Scotland. *Wildfowl* 44: 111-120.

Durell, S.E.A. Le V., Goss-Custard, J.D. & Caldow, R.W.G. 1993. Sex related differences in diet and feeding method in the Oystercatcher *Haematopus astralegus*. *J. Anim. Ecol.* 62: 205-215.

Forshaw, W.D. 1993. Report on wild geese and swans in Lancashire 1992-93. Unpubl. report, 10pp.

Fox, A.D. 1993a. *Report of the 1992-93 national census of Greenland White-fronted Geese in Britain.* Greenland White-fronted Goose Study, 11pp.

Fox, A.D. 1993b. Pintail. In: Gibbons, D.W., Reid, J.B. & Chapman, R. (Eds.) *The New Atlas of Breeding Birds in Britain and Ireland: 1988-1991.* Poyser, London.

Fox, A.D. & Meek, E.R. 1993. History of the Northern Pintail breeding in Britain and Ireland. *Brit. Birds* 86: 151-162.

Gibbons, D.W., Reid, J.B. & Chapman, R. (Eds.) 1993. *The New Atlas of Breeding Birds in Britain and Ireland: 1988-1991.* Poyser, London.

Gilburn, A.S. & Kirby, J.S. 1992. Winter status, distribution and habitat use by Teal in the United Kingdom. WWT report to JNCC, Slimbridge, 21pp.

Green, M. & Elliott, D. 1993. Surveys of wintering birds and cetaceans in northern Cardigan Bay, 1990-93. Unpublished report, Friends of Cardigan Bay, 32pp.

Green, R.E. & Robins, M. 1993. The decline of the ornithological importance of the Somerset Levels and Moors, England, and changes in the management of water levels. *Biol. Conserv.* 66: 95-106.

Greenwood, J.J.D. & Delany, S.N. in press. A method for estimating Mute Swan breeding populations in Great Britain in three years. *Proc. 12th IBCC/EOAC Conf.*

Gudmundsson, G.A. 1993. The spring migration pattern of arctic birds in southwest Iceland, as recorded by radar. *Ibis* 135: 166-176.

Gudmundsson, G.A. & Lindstrom, A. 1992. Spring migration of Sanderlings *Calidris alba* through SW Iceland: wherefrom and whereto? *Ardea* 80: 315-326.

Harradine, J. (Ed.) 1991. Canada Geese - problems and management needs. *Proc. BASC Canada Goose Conference.*

Henderson, I.G., Peach, W.J., & Baillie, S.R. 1993. The hunting of Snipe and Woodcock in Europe: a ringing recovery analysis. *BTO Research Report No. 115.* BTO, Thetford.

Hill, D. & Player, A. 1992. Behavioural responses of Black-headed Gulls and Avocets to two methods of control of gull productivity. *Bird Study* 39: 34-42.

Hill, D., Underhill, M.C., & Robinthwaite, J. 1993. Ornithological survey and assessment of core SPA waterbodies: southwest London (in respect of widening the M25, J 12-25). Final Report. Wetlands Advisory Service and Ecoscope Applied Ecologists report to the Department of Transport, 180pp.

Hogan-Warburg, A.J. 1992. Female choice and the evolution of mating strategies in the Ruff *Philomachus pugnax* (L.). *Ardea* 80: 395-403.

Hoglund, J., Montgomerie, R., & Widerno, F. 1993. Costs and consequences of variation in size of Ruff leks. *Behav. Ecol. Sociobiol.* 32: 31-39.

Holmgren, N., Ellegren, H. & Pettersson, J. 1993. Stopover length, body mass and fuel deposition rate in autumn migrating adult dunlins *Calidris alpina*: evaluating the effects of moulting status and age. *Ardea* 81: 9-20.

Hötker, H. (Ed.) 1991. Waders breeding on wet grassland. *Wader Study Group Bull.* 61. Supplement, 107 pp.

del Hoyo, J., Elliott, A. & Sargatal, J. (Eds.) 1992. *Handbook of the Birds of the World.* Vol. 1, Lynx Edicions, Barcelona.

Hughes, B. & Grussu, M. in press. The Ruddy Duck in Europe and the threat to the White-headed Duck. In: Andrews, J. (Ed.) *Britain's Birds in 1991-92: The Conservation and Monitoring Review.*

Hutchinson, C.D. 1979. *Ireland's wetlands and their birds.* Dublin.

Jones, T.A. 1989. Shelduck in the Severn Estuary, May-August 1989. WWT internal report, Slimbridge.

Kirby, J.S. 1993. Position Statement concerning Cormorant Research, Conservation and Management, Gdansk 1993. WWT, Slimbridge, 4pp.

Kirby, J.S. & Mitchell, C. 1993. Distribution and status of wintering Shovelers *Anas clypeata* in Great Britain. *Bird Study* 40: 170-180.

Kirby, J.S. & Owen, M. 1993. Woolston Eyes - past, present but no future? Pp. 20-22. In: Andrews, J. & Carter, S.P. (Eds). *Britain's Birds in 1990-91: the Conservation and Monitoring Review.* BTO/JNCC, Thetford.

Kirby, J.S. & Sellers, R.M. in press. Recent trends in the numbers and distribution of Cormorants in Great Britain. *Proc. 3rd International Cormorant Research Group Meeting, Gdansk, 1993.*

Kirby, J.S., Evans, R.J. & Fox, A.D. 1993. Wintering seaducks in Britain and Ireland: populations, threats, conservation and research priorities. *Aquatic Conservation: Marine and Freshwater Ecosystems* 3: 105-137.

Kirby, J.S., Gilburn, A.S. & Sellers, R.M. in press, b. Status, distribution and habitat use by Cormorants *Phalacrocorax carbo* wintering in Britain. *Ardea.*

Kirby, J.S., Holmes, J. & Sellers, R.M. in press, c. Conservation and management of Cormorants in Great Britain: the current situation. *Proc. 3rd International Cormorant Research Group Meeting, Gdansk, 1993.*

Kirby, J.S., Rees, E.C., Merne, O.J. & Gardarsson, A. 1992. International census of Whooper Swans *Cygnus cygnus* in Britain, Ireland and Iceland: January 1991. *Wildfowl* 43: 20-26.

Kirby, J.S., Salmon, D.G. & Atkinson-Willes, G.L. in press, a. Index numbers for waterbird populations. III. Long-term trends in the abundance of wintering wildfowl in Great Britain, 1966/67 - 1991/2. *J. Appl. Ecol.*

Lack, P. (Ed.) 1986. *The Atlas of Wintering Birds in Britain and Ireland.* Poyser, Calton.

Lambeck, R.H.D. & Wessel, E.G.J. 1993. A note on Oystercatchers from the Varangerfjord, NE Norway. *Wader Study Group Bull.* 66: 74-79.

Larsen, T. 1993. Information parasitism in foraging Bar-tailed Godwits *Limosa lapponica. Ibis* 135: 271-276.

Marquiss, M. 1993. Grey Heron. In: Gibbons, D.W., Reid, J.B. & Chapman, R. (Eds.) *The New Atlas of Breeding Birds in Britain and Ireland: 1988-1991.* Poyser, London.

Marquiss, M. & Duncan, K. 1993. Variation in the abundance of Red-breasted Mergansers *Mergus serrator* on a Scottish river in relation to season, year, river hydrography, salmon density and spring culling. *Ibis* 135: 33-41.

Martin, A.P., Uttley, J.D. & Underhill, L.G. 1992. An unfaithful Curlew Sandpiper? *Wader Study Group Bull.* 66: 41-42.

Mead, C. 1993. Kingfisher. In: Gibbons, D.W., Reid, J.B. & Chapman, R. (Eds.) *The New Atlas of Breeding Birds in Britain and Ireland: 1988-1991*. Poyser, London.

Mead, C.J. & Clark, J.A. 1993. Report on Bird Ringing in Britain and Ireland for 1991. *Ringing and Migration* 4: 1-72.

Mead, C.J., Clark, J.A. & Peach, W.J. 1993. Report on Bird Ringing in Britain and Ireland for 1992. *Ringing & Migration* 14: 152-200.

Meek, E.R. 1993. The status of the Pintail in the Orkney Islands. *Scott. Birds* 17: 14-19.

Meltofte, H. 1993. *Vadefugletraekket Gennem Danmark*. Zoological Museum, Kobenhauns Universitet.

Mitchell, C. & Cranswick, P.A. 1993. The 1992 national census of Pink-footed and Greylag Geese in Britain. WWT report to JNCC, Slimbridge, 13pp.

Mitchell, C., Boyer, P.R., Shimmings, P. & Delany, S. 1993. Greylag Geese on the Uists 1992-93. WWT report to JNCC, Slimbridge, 8pp.

Monval, J.-Y. & Pirot, J.-P. 1989. Results of the IWRB International Waterfowl Census 1967-86. *IWRB Spec. Publ.* 8, IWRB, Slimbridge.

Moser, M.E. 1987. A revision of population estimates for waders *(Charadrii)* wintering on the coastline of Britain. *Biol. Conserv.* 39: 153-164.

Moss, D. 1993. Little Grebe. In: Gibbons, D.W., Reid, J.B. & Chapman, R. (Eds.) *The New Atlas of Breeding Birds in Britain and Ireland: 1988-1991*. Poyser, London.

Moss, D. & Moss, G.M. 1993. Breeding biology of the Little Grebe in Britain and Ireland. *Bird Study* 40: 107-114.

OAG Münster & OAG Schleswig-Holstein. 1992. (Numbers of Ruffs during autumn migration in Germany 1990.) *Vogelwelt* 113: 102-113. In German, with English summary.

O'Brien, M. & Smith, K.W. 1992. Changes in the status of waders breeding on wet lowland grasslands in England and Wales between 1982 and 1989. *Bird Study* 39: 165-176.

Ogilvie M.A. 1977. The number of Canada Geese in Britain, 1976. *Wildfowl* 28: 27-34.

Owen, M., Atkinson-Willes, G.L. & Salmon, D.G. 1986. *Wildfowl in Great Britain*. 2nd Edition. University Press, Cambridge.

Owen, M. & Black, J.M. 1991. Geese and their future fortune. *Ibis* 133: 28-35.

Parr, R. 1993. Nest predation and numbers of Golden Plovers *Pluvialis apricaria* and other moorland waders. *Bird Study* 40: 223-231.

Phillips, V.E. & Wright, R.M. 1993. The difference in behaviour and feeding success of tame mallard ducklings *Anas platyrhynchos* in the presence of high and low fish populations at a gravel pit site, with reference to wild brood distribution. *Wildfowl* 44: 69-74.

Pickering, S.P.C. 1993. Breeding waterfowl in the Cotswold Water Park. WWT unpubl. report, 9pp.

Piotrowski, S.H. (Ed.) 1993. *Suffolk Birds 1992*. Suffolk Naturalists' Society, Ipswich.

Prater, A.J. 1981. *Estuary Birds of Britain and Ireland*. Poyser, Calton.

Prŷs-Jones, R.P., Underhill, L.G. & Waters, R.J. in press. Index numbers for waterbird populations. II. Coastal wintering waders in the United Kingdom, 1970/71 - 1990/91. *J. Appl. Ecol.*

Pulliainen, E. & Saari, L. 1993. Breeding biology of the Whimbrel *Numenius phaeopus* in eastern Finnish Lapland. *Ornis Fenn.* 70: 110-116.

Quinn, J.L., Still, L., Carrier, M. & Lambdon, P. 1993. The distribution of waterfowl and other bird species on the Solway Firth, at Chapelcross and in the Annan catchment: October 1991 - July 1993, Volumes 1-3. Wetlands Advisory Service report to British Nuclear Fuels Ltd.

Ratcliffe, D.A. (Ed.) 1977. *A Nature Conservation Review*. Cambridge University Press, Cambridge.

Reay, P. 1993. *The Tamar and Exe Avocets*. Supplement 5 (1992/93). Department of Biological Sciences, Polytechnic South West.

Rees, E.C., Bowler, J.M., Beekmann, J.H., Andersen-Harild, P., Mineyev, Yu N., Shchladilov, Yu M., Belousova, A.V., Morozov, Yu V., Poot, M., Peberdy, K.J. & Scott, D.K. in prep. International collaborative study of Bewick's Swans nesting in the European Northeast of Russia. *Russian Journal of Ornithology*.

Rehfisch, M.M., Langston, R.H.W., Clark, N.A. & Forrest, C. 1993. A guide to the provision of refuges for waders. An analysis of Wash Wader Ringing Group data. *BTO Research Report No. 120*.

Rose, P.M. (Ed.) 1993. *Ruddy Duck (Oxyura jamaicensis) European Status Report 1993*. IWRB, 19pp.

Rose, P.M. & Taylor, V. 1993. *Western Palaearctic and South West Asia Waterfowl Census 1993*. IWRB, Slimbridge.

Rose, P.M. & Scott, D.A. 1994. Waterfowl population estimates. *IWRB/AWB/WA Spec. Publ. 18*, IWRB, Slimbridge.

Salmon, D.G. 1986. Shoveler. In: Lack, P. (Ed.) *The Atlas of Wintering Birds in Britain and Ireland*. Poyser, Calton.

Salmon, D.G. 1988. A late summer survey of Britain's wildfowl. *Wildfowl* 39: 155-163.

Serra, L., Baccetti, N. & Magnani, A. 1992. Migration, wintering and moult of Little Stints in northeast Italy. *Wader Study Group Bull.* 64: 9.

Sezkely, T. & Bamerger, Z. 1992. Predation of waders *(Charadrii)* on prey populations: an exclosure experiment. *Wader Study Group Bull.* 64: 14.

Sharrock, J.T.R. 1976. *The Atlas of Breeding Birds in Britain and Ireland*. Poyser, Berkhamstead.

Shepherd, M. & Clark, N.A. 1993. The effect of commercial cockling on the numbers of wintering waterfowl on the Solway estuary. *BTO Research Report No. 128*.

Sheppard, R. in prep. *Ireland's wetland wealth: the birdlife of the estuaries, lakes, coasts, rivers, bogs and turloughs of Ireland. The report of the winter wetlands survey, 1984/85 to 1986/87.* Irish Wildbird Conservancy, Dublin.

Shimmings, P., Choudhury, S., Owen, M. & Black, J.M. 1993. Wintering Barnacle Geese on the Solway Firth, 1992-93. WWT report to Scottish Natural Heritage, Slimbridge.

Smit, C.J. & Piersma, T. 1989. Numbers, midwinter distribution and migration of wader populations using the East Atlantic flyway. Pp 24-64. In: Boyd, H. & Pirot, J-Y. (Eds.) *Flyways and reserve networks for waterbirds.* IWRB Spec. Publ. 9, Slimbridge.

Smith, K.W., Reed, J.M. & Trevis, B.E. 1992. Habitat use and site fidelity of Green Sandpipers *Tringa ochropus* wintering in Southern England. *Bird Study* 19: 155-164.

Summers, R.W., Underhill, L.G., Nicoll, M. Rae, R. & Piersma, T. 1992. Seasonal, size- and age-related patterns in body-mass and composition of Purple Sandpipers *Calidris maritima* in Britain. *Ibis* 134: 346-354.

Suter, W. & Van Eerden, M.R. 1992. Simultaneous mass starvation of wintering diving ducks in Switzerland and the Netherlands: a wrong decision in the right strategy? *Ardea* 80: 229-242.

Taylor, K. 1986. Kingfisher. In: Lack, P. (Ed). *The Atlas of Wintering Birds in Britain and Ireland.* Poyser, Calton.

Thompson, P.S. & Hale, W.G. 1993. Adult survival and numbers in a coastal breeding population of Redshank *Tringa totanus* in northwest England. *Ibis* 135: 61-69.

Toomer, D.K. & Clark, N.A. 1993. The roosting behaviour of waders and wildfowl in Cardiff Bay: Winter 1992/93. *BTO Research Report No. 116.*

Trolliet, B., Girard, O., Fouquet, M., Ibañez, F., Triplet, P. & Léger, F. 1992. L'effectif de Combattants *Philomachus pugnax* hivernant dans le Delta du Sénégal. *Alauda* 60: 159-163. In French, with English summary.

Turpie, J.K. & Hockey, P.A.R. 1993. Comparative diurnal and nocturnal foraging behaviour and energy intake of premigratory Grey Plovers *Pluvialis squatarola* and Whimbrels *Numenius phaeopus* in South Africa. Ibis 135: 156-165.

Underhill, L.G. 1989. Indices for waterbird populations. *BTO Research Report 52.*

Underhill, L.G. & Prŷs-Jones, R. in press. Index numbers for waterbird populations. I. Review and methodology. *J. Appl. Ecol.*

Underhill, M.C. 1993. The distribution and abundance of Cormorants and Goosanders on the River Wye and its main tributaries in 1993. Wetlands Advisory Service report to the National Rivers Authority (Welsh Region) and the Countryside Council for Wales, 102pp.

Underhill, M.C., & Robinthwaite, J. 1993a. Waterfowl use, management and human usage of southwest London waterbodies: 1987-88 to 1992-93. Wetlands Advisory Service report to Thames Water plc and English Nature, 243pp.

Underhill, M.C., & Robinthwaite, J. 1993b. A survey of wintering and breeding birds of the Wraysbury gravel pit complex. Wetlands Advisory Service report to W.S. Atkins Environment, Surrey, 166pp.

Underhill, M.C., Kirby, J.S., Bell, M.C & Robinthwaite, J. 1993. Use of waterbodies in southwest London by waterfowl: an investigation into factors affecting distribution, abundance and community structure. Wetlands Advisory Service report to Thames Water plc and English Nature, 138pp.

Van Rhijn, J.G. 1991. *The Ruff: individuality in a gregarious wading bird.* Poyser, London.

Waters, R.J. in prep. A guide to Birds of Estuaries Enquiry counting procedure at non-estuarine sites. *BTO Research Report No. 110.*

Way, L.S., Grice, P., MacKay, A., Galbraith, C.A., Stroud, D.A. & Pienkowski, M.W. 1993. *Ireland's internationally important bird sites: a review of sites for the EC Special Protection Area network.* JNCC, Peterborough.

Winfield, I.J., Winfield, D.K. & Tobin, C.M. (1992) Interactions between the roach, *Rutilus rutilus,* and waterfowl populations of Lough Neagh, Northern Ireland. *Environ. Biol. of Fishes* 33: 207-214.

Appendix I. INTERNATIONAL AND NATIONAL IMPORTANCE

Criteria for International Importance have been agreed by the Contracting Parties to the Ramsar Convention on Wetlands of International Importance (Ramsar Convention Bureau 1988). Under one criterion, a wetland is considered Internationally Important if it regularly holds 1% of the individuals in a population of one species or subspecies of waterfowl, while any site regularly holding a total of 20,000 waterfowl also qualifies. Britain and Ireland's wildfowl belong to the north-west European population (Pirot *et al.* 1989), and the waders to the east Atlantic flyway population (Smit & Piersma 1989). A wetland in Britain is considered Nationally Important if it regularly holds 1% of the estimated British population of one species or subspecies of waterfowl, and in Northern Ireland important in an all-Ireland context if it holds 1% of the estimated all-Ireland population (see Table 56).

Since *Wildfowl and Wader Counts 1991-92*, a further 11 Ramsar sites and 15 SPAs have been designated in the UK, some sites receiving dual designation.

Ramsar designation only
Roydon Common (Norfolk)
Crymlyn Bog (W Glam)
Malham Tarn (N Yorks)

SPA classification only
Sheep Island (Co Antrim)
Great Yarmouth North Denes (Norfolk)
Ouse Washes (Cambs/Norfolk)
Hornsea Mere (Humbs)
Flamborough Head & Bempton Cliffs (Humbs/N Yorks)
Salisbury Plain (Wilts/Hants)
Bowland Fells (Lancs)

SPA and Ramsar designation
Nene Washes (Cambs)
Gibraltar Point/The Wash Phase 2 (Lincs)
South Tayside Goose Roosts (Tayside)
Hamford Water (Essex)
Lower Derwent Valley (Humbs/N Yorks)
The New Forest (Hants)
Medway Estuary and Marshes (Kent)
Stodmarsh (Kent)

These designations represent real progress in the designation programme (see *Conservation and Management*). In total, 69 Ramsar sites and 77 SPAs have been designated in the UK.

(R) = Ramsar site only; (S) = SPA only; the remainder have dual designation.

Abberton Reservoir
Abernethy Forest (S)
Ailsa Craig (S)
Alt Estuary
Bowland Fells (S)
Bridgend Flats
Bridgwater Bay (R)
Bure Marshes (R)
Burry Inlet
Cairngorm Lochs (R)
Chesil Beach/Fleet
Chew Valley Lake (S)
Chichester/Langstone Harbours
Chippenham Fen (R)
Claish Moss (R)
Coquet Island (S)
Cors Caron (R)
Cors Fochno/Dyfi (R)
Crymlyn Bog (R)
Dee Estuary
Derwent Ings
Eilean na Muice Duibhe
  (Duich Moss)
Esthwaite Water (R)
Exe Estuary
Fala Flow
Farne Islands (S)
Feur Lochain
Flamborough Head
  & Bempton Cliffs (S)

Flannan Isles (S)
Forth Islands (S)
Fowlsheugh (S)
Gibraltar Point/The Wash Phase 2
Glac-na-Criche
Gladhouse Reservoir
Glannau Aberdaron (S)
Glannau Ynys Gybi (S)
Grassholm (S)
Great Yarmouth North Denes (S)
Gruinart Flats
Hamford Water
Handa Island (S)
Hickling Broad/Horsey Mere (R)
Holburn Lake and Moss
Hornsea Mere (S)
Hoselaw Loch
Irtinghead Mires (R)
Laggan Peninsula (S)
Leighton Moss
Lindisfarne
Llyn Idwal (R)
Llyn Tegid (R)
Loch An Duin (R)
Lochs
Druidibeg/a'Machair/Stillgary
Loch Eye
Loch Ken/Dee Marshes
Loch Leven (R)
Loch Lomond (R)

Loch of Lintrathen
Loch of Skene
Loch Spynie
Loughs Neagh/Beg (R)
Lower Derwent Valley
Malham Tarn (R)
Martin Mere
Medway Estuary and Marshes
Minsmere/Walberswick
Moor House (S)
Nene Washes
North Norfolk Coast
Old Hall Marshes
Orfordness/Havergate (S)
Ouse Washes
Pagham Harbour
Porton Down (S)
Priest Island (S)
Rannoch Moor (R)
Redgrave and South
  Lopham Fens (R)
Rhum (S)
Ribble Estuary (part) (S)
Rockcliffe Marshes
Rostherne Mere (R)
Roydon Common (R)
Rutland Water
Salisbury Plain (S)
Sheep Island (S)
Shiant Isles (S)

Silver Flowe (R)
Skokholm and Skomer Islands (S)
South Tayside Goose Roosts
St Kilda (S)
Stodmarsh
Swan Island (S)
The New Forest
The Wash
The Swale
Traeth Lafan (S)
Upper Severn Estuary
Upper Solway
Walmore Common
Ynys Feurig (S)

1% Levels for National and international importance

A wetland is considered important in a national or all-Ireland context if it regularly holds 1% of one species, sub-species or population of waterfowl in Great Britain or the island of Ireland respectively. Similarly, a wetland is of international importance if it supports 1% of the international population. Many wildfowl wintering in Britain and Ireland form part of the North-West European population, whilst many waders form part of populations that may range over much of the East Atlantic. Table 56 lists the numbers of each species that represent 1% of the British, all-Ireland and international waterfowl populations where known. Thus, any site regularly supporting this number of birds potentially qualifies for designation under national legislation or international directives or conventions. The international population ranges for each species and sub-species are also given in the table. However, it should be noted that, where 1% of the national population is less than 50 birds, 50 is normally used as a minimum qualifying level for the designation of sites of national importance. 1% levels have not been derived for introduced species since these species are not included in the relevant parts of the legislation and important sites (e.g. SSSIs) would not be identified on the basis of numbers of these birds. Sources of qualifying levels represent the most up-to-date figures following recent reviews: for British wildfowl see Kirby (in prep.); for British waders see Cayford & Waters (in prep.); for all-Ireland importance for divers see Danielsen *et al.* (1993) and for other waterfowl see Whilde (in prep.) cited in Way *et al.* (1993). Following a recent workshop in Denmark on international populations, international criteria follow Smit & Piersma (1989) or Rose & Scott (1994). Several of the international populations are expected to be revised shortly (P. Rose pers. comm.).

Table 56. 1% LEVELS FOR NATIONAL AND INTERNATIONAL IMPORTANCE

	Great Britain	all-Ireland	International	Population
Red-throated Diver	50	10 *	750	Europe/Greenland
Black-throated Diver	7 *	1 *	1,200	Europe/W Siberia
Great Northern Diver	30 *	?	50	Europe
Little Grebe	30 *	?	?	W Palaearctic
Great Crested Grebe	100	30 *	?	NW Europe
Red-necked Grebe	1 *	?	300	NW Europe
Slavonian Grebe	4 *	?	50	NW Europe
Black-necked Grebe	1 *	?	1,000	W Palaearctic
Cormorant	130	?	1,200	NW Europe
Grey Heron	?	?	4,500	Europe/N Africa
Mute Swan	260	55	1,800	NW Europe
Bewick's Swan	70	25 *	170	Europe (wintering)
Whooper Swan	55	100	170	Iceland
Bean Goose	4 *	+ *	800	W Tundra
Pink-footed Goose: Iceland/Greenland	1,900	+ *	1,900	Iceland/Greenland
European White-fronted Goose	60	+ *	4,500	NW Europe
Greenland White-fronted Goose	140	140	260	Greenland
Greylag Goose: Iceland	1,000	40 *	1,000	Iceland
Hebrides/N Scotland	50	n/a	50	Scotland
Barnacle Goose: Greenland	270	753	20	Greenland
Svalbard	120	+ *	120	Svalbard
Dark-bellied Brent Goose	1,000	+ *	2,500	Siberia
Light-bellied Brent Goose: Canada/Greenland	+ *	200	200	Canada/Greenland
Svalbard	25 *	+ *	40	Svalbard
Shelduck	750	70	2,500	NW Europe
Wigeon	2,800	1,250	7,500	NW Europe
Gadwall	80	+ *	250	NW Europe
Teal	1,400	650	4,000	NW Europe
Mallard	5,000	500	20,000 **	NW Europe
Pintail	280	60	700	NW Europe
Garganey	+ *	+ *	20,000 **	W Africa (wintering)
Shoveler	100	65	400	NW Europe
Red-crested Pochard	+ *	+ *	200	SW/Central Europe
Pochard	440	400	3,500	NW Europe
Tufted Duck	600	400	7,500	NW Europe
Scaup	110	30 *	3,100	NW Europe
Eider	750	20 *	20,000 **	Europe
Long-tailed Duck	230	+ *	20,000 **	Iceland/Greenland

	Great Britain	all-Ireland	International	Population
Common Scoter	230	40 *	8,000	NW Europe
Velvet Scoter	30 *	+ *	2,500	NW Europe
Goldeneye	170	110	3,000	NW Europe
Smew	2 *	+ *	150	NW Europe
Red-breasted Merganser	100	20 *	1,000	NW Europe
Goosander	90	+ *	1,500	NW Europe
Coot	1,100	250	15,000	NW Europe
Oystercatcher	3,600	500	9,000	Europe/W Africa (wintering)
Avocet	10 *	+ *	700	Europe/NW Africa (breeding)
Ringed Plover	290	125	500	Europe/NW Africa (wintering)
Golden Plover	2,500	2,000	18,000	NW Europe (breeding)
Grey Plover	440	40 *	1,500	E Atlantic
Lapwing	20,000 **	2,500	20,000 **	Europe/W Africa
Knot C. c. canutus	2,900	375	3,500	W Europe/Canada
C. c. islandica			5,000	W Africa/W Siberia
Sanderling	230	35 *	1,000	E Atlantic
passage	300			
Purple Sandpiper	210	10 *	500	E Atlantic
Dunlin C. a. arctica			150	Greenland (breeding)
C. a. schinzii (Icelandic)			8,000	Iceland/Greenland (breeding)
C. a. schinzii (temperate)			200	UK/Ireland/Baltic
C. a. alpina	5,300	1,250	14,000	Europe (breeding)
Ruff	7 *	+ *	?	W Africa (wintering)
Jack Snipe	?	250	?	Europe/W Africa (wintering)
Snipe	?	?	10,000	Europe/W Africa (breeding)
Black-tailed Godwit	75	90	700	Iceland (breeding)
Bar-tailed Godwit	500	175	1,000	W Europe (wintering)
Whimbrel	+ *	+ *	6,500	Europe/W Africa (wintering)
passage	50			
Curlew	1,200	875	3,500	Europe/NW Africa
Spotted Redshank	+ *	+ *	1,500	Europe/W Africa
Redshank T. t. totanus	1,100	245	1,500	Europe/W Africa (wintering)
T. t. robusta			1,500	NW Europe (wintering)
Greenshank	+ *	9 *	3,000	Europe/W Africa
Turnstone	650	225	700	Europe (wintering)

? Population size not accurately known.
+ Population too small for meaningful figure to be obtained.
* Where 1% of the British or all-Ireland wintering population is less than 50 birds, 50 is normally used as a minimum qualifying level for national or all-Ireland importance respectively.
** A site regularly holding more than 20,000 waterfowl qualifies as internationally important by virtue of absolute numbers.

Appendix 2. LOCATIONS OF WeBS COUNT SITES

The location of all count sites or areas mentioned in this booklet are given here. The location of estuaries are given in Figure 2, whilst inland sites are listed in Table 57 in alphabetical order, with the 1 km square grid reference for the centre of the area and the county or district.

Figure 2. MAP OF THE BRITISH ISLES SHOWING THE LOCATIONS OF ALL ESTUARIES CONSIDERED IN THIS REPORT

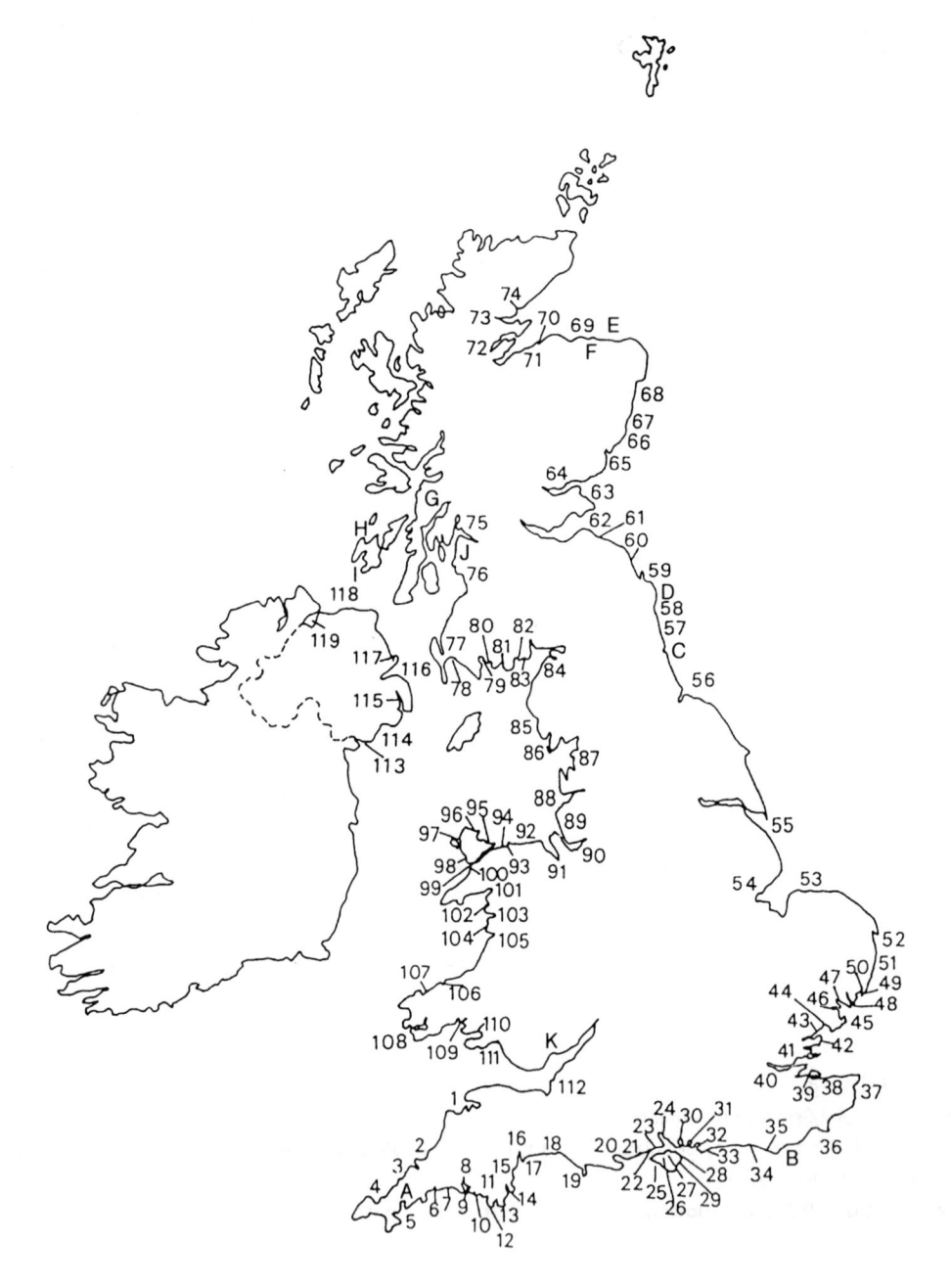

Site numbers are as follows: 1 Taw/Torridge; 2 Camel; 3 Gannel; 4 Hayle; 5 Fal complex; 6 Fowey; 7 Looe; 8 Tamar; complex; 9 Plym; 10 Yealm; 11 Erme; 12 Avon; 13 Kingsbridge; 14 Dart; 15 Teign; 16 Exe; 17 Otter; 18 Axe; 19 The Fleet/Wey; 20 Poole Harbour; 21 Christchurch Harbour; 22 NW Solent; 23 Beaulieu; 24 Southampton Water; 25 Yar; 26 Newtown; 27 Medina; 28 Wootton; 29 Brading Harbour; 30 Portsmouth Harbour; 31 Langstone Harbour; 32 Chichester Harbour; 33 Pagham Harbour; 34 Adur; 35 Newhaven; 36 Rye Harbour/Pett Levels; 37 Pegwell Bay; 38 Swale; 39 Medway; 40 Thames; 41 Crouch/Roach; 42 Dengie; 43 Blackwater; 44 Colne; 45 Hamford Water; 46 Stour; 47 Orwell; 48 Deben; 49/50 Alde complex; 51 Blyth; 52 Breydon Water; 53 N Norfolk Marshes; 54 Wash; 55 Humber; 56 Tees; 57 Blyth; 58 Coquet; 59 Lindisfarne; 60 Tweed; 61 Tyninghame; 62 Forth; 63 Eden; 64 Tay; 65 Montrose Basin; 66 Dee; 67 Don; 68 Ythan; 69 Spey; 70/71 Inner Moray Firth; 72 Cromarty Firth; 73 Dornoch Firth; 74 Loch Fleet; 75 Inner Clyde; 76 Irvine; 77 Loch Ryan; 78 Luce Bay; 79 Wigtown Bay; 80 Fleet Bay; 81 Kirkcudbright Bay; 82 Auchencairn Bay; 83 Rough Firth; 84 Solway; 85 Irt/Mite/Esk; 86 Duddon; 87 Morecambe Bay; 88 Ribble; 89 Alt; 90 Mersey; 91 Dee; 92 Clwyd; 93 Conwy; 94 Lavan Sands; 95 Red Wharf Bay; 96 Dulas Bay; 97 Inland Sea; 98 Cefni; 99 Braint; 100 Foryd Bay; 101 Traeth Bach; 102 Artro; 103 Mawddach; 104 Dysynni; 105 Dyfi; 106 Teifi; 107 Nyfer; 108 Cleddau; 109 Carmarthen Bay; 110 Burry; 111 Swansea Bay; 112 Severn; 113 Carlingford Lough; 114 Dundrum Bay; 115 Strangford Lough; 116 Belfast Lough; 117 Lough Larne; 118 Bann; 119 Lough Foyle. A Helford; B Cuckmere; C Tyne; D South Alnmouth; E Banff; F Lossie; G Loch Gilp; H Loch Gruinart (Islay); I Loch Indaal (Islay); J Hunterston; K Afan; L Ogmore.

Table 57. THE LOCATION OF INLAND WeBS SITES

Site	1 km square	County
Abberton Reservoir	NT 4581	Lothian
Abbots Moss	NY 5142	Cumbria
Aird Meadow	NS 3658	Strathclyde
Alaw Reservoir	SH 3968	Gwynedd
Aller Moor	ST 3929	Somerset
Alton Water	TM 1536	Suffolk
Alvecote Pools	SK 2504	Warwickshire
Annaghroe	H 7344	Tyrone
Ancum Loch	HY 7654	Orkney
Appin/Eriska/Benderloch	NM 9138	Strathclyde
Ardleigh Reservoir	TM 0328	Essex
Ardoch Loch	NN 8408	Tayside
Arlesford Pond	SU 5933	Hampshire
Arundel WWT	TQ 0207	West Sussex
Bardolf Water Meadows	ST 7795	Dorset
Barn Elms Reservoirs	TQ 2277	Greater London
Belvide Reservoir	SJ 8610	Staffordshire
Benacre Broad	TM 5383	Suffolk
Berney Marshes	TG 4605	Norfolk
Besthorpe/Girton Gravel Pits	SK 8165	Nottinghamshire
Bewl Water	TQ 6733	East Sussex
Blagdon Lake	ST 5150	Avon
Blenheim Park Lake	SP 4316	Oxfordshire
Blickling Lake	TG 1729	Norfolk
Blithfield Reservoir	SK 0524	Staffordshire
Borth/Ynyslas	SN 6092	Dyfed
Bosherston Lake	SR 9794	Dyfed
Brandon Grounds	SP 4176	Warwickshire
Broad Bay	NB 4733	Western Isles
Broomhill Flash	SE 4102	South Yorkshire
Buckden/Stirtloe Gravel Pits	TL 2066	Cambridgeshire
Burham Marsh	TQ 7362	Kent
Caban Coch Reservoir	SN 9163	Powys
Caistron Quarry	NU 0001	Northumberland
Cameron Reservoir	NO 4711	Fife
Carron Valley Reservoir	NS 6884	Central
Carsebreck/Rhynd Lochs	NN 8609	Tayside
Castle Howard Lake	SE 7170	North Yorkshire
Castle Loch, Lochmaben	NY 0881	Dumfries & Galloway
Catcott Heath	ST 4041	Somerset
Cheddar Reservoir	ST 4454	Somerset
Cheshunt Gravel Pit	TL 3602	Hertfordshire
Chew Valley Lake	ST 5659	Avon
Chichester Gravel Pits	SU 8703	West Sussex
Clifford Hill Gravel Pit	SP 8061	Northamptonshire
Cloddach Gravel Pit	NJ 2059	Highland
Clumber Park Lake	SK 6347	Nottinghamshire
Colliford Reservoir	SX 1871	Cornwall
Cotswold Water Park East	SU 1999	Gloucestershire/Oxfordshire
Cotswold Water Park West	SU 0595	Gloucestershire/Wiltshire
Cowgill Reservoirs	NT 0327	Strathclyde
Cresswell Ponds	NZ 2993	Northumberland
Crombie Loch	NO 5240	Tayside
Danna/Keils Peninsula	NR 7383	Strathclyde
Daventry Reservoir	SP 5763	Northamptonshire
Dinnet Lochs	NJ 4800	Grampian
Dorchester Gravel Pits	SU 5795	Oxfordshire
Doxey Marshes	SJ 9024	Staffordshire
Drakelow Gravel Pit	SK 2320	Derbyshire
Drift Reservoir	SW 4328	Cornwall

Site	1 km square	County
Drummond Pond	NN 8518	Tayside
Dungeness	TR 0619	Kent
Dupplin Loch	NO 0320	Tayside
Durleigh Reservoir	ST 2636	Somerset
Eccup Reservoir	SE 2941	West Yorkshire
Eglwys Nunydd Reservoir	SS 7984	West Glamorgan
Endrick Mouth, Loch Lomond	NS 4388	Strathclyde
Essenside Loch	NT 4520	Borders
Eyebrook Reservoir	SP 8595	Leicestershire
Fairburn Ings	SE 4627	North Yorkshire
Fala Flow	NT 4258	Lothian
Farmoor Reservoirs	SP 4406	Oxfordshire
Farmwood Pool	SJ 8173	Cheshire
Fedderate Reservoir	NJ 8652	Grampian
Fen Drayton Gravel Pits	TL 3470	Cambridgeshire
Fiddlers Ferry Lagoons	SJ 5585	Cheshire
Fleet Pond	SU 8255	Surrey
Frensham Ponds	SU 8440	Surrey
Gladhouse Reservoir	NT 2953	Lothian
Grafham Water	TL 1568	Cambridgeshire
Gunthorpe Gravel Pits	SK 6744	Nottinghamshire
Gunton Parks	TG 2234	Norfolk
Haddo House Lakes	NJ 8734	Grampian
Hallington Reservoir	NY 9776	Northumberland
Hamilton Low Parks	NS 7257	Strathclyde
Hamner Mere	SJ 4539	Clwyd
Hanningfield Reservoir	TQ 7398	Essex
Hardley Flood	TM 3899	Norfolk
Hay-a-Park Gravel Pits	SE 3658	North Yorkshire
Heaton Park Reservoir	SD 8205	Greater Manchester
Hickling Broad	TG 4121	Norfolk
Hilfield Park Reservoir	TQ 1595	Hertfordshire
Hirsel Lake	NT 8240	Borders
Holburn Lake and Moss	NU 0536	Northumberland
Holden Wood Reservoir	SD 7722	Lancashire
Holme Pierrepoint Gravel Pits	SK 6239	Nottinghamshire
Holywell Pond	NZ 3175	Northumberland
Horsey Mere	TG 4415	Norfolk
Hoselaw Loch	NT 8031	Borders
Hoveringham/Bleasby Gravel Pits	SK 7047	Nottinghamshire
Hule Moss	NT 7149	Borders
Kedleston Park	SK 3141	Derbyshire
Kilconquhar Loch	NO 4801	Fife
King George V Reservoir	TQ 3796	Greater London
Kingsbury Water Park/Coton Pools	SP 2096	Warwickshire
Kings Mill Reservoir	SK 5159	Nottinghamshire
Kinmount Ponds	NY 1468	Dumfries & Galloway
Lackford Gravel Pits	TL 7971	Suffolk
Lake of Menteith	NN 5700	Central
Lancaster Canal	SD 4766	Lancashire
Leighton Moss	SD 4875	Lancashire
Leighton/Roundhill Reservoirs	SE 1678	North Yorkshire
Leybourne/New Hythe Gravel Pits	TQ 6959	Kent
Liddel Loch	ND 4583	Orkney
Little Paxton Gravel Pits	TL 1963	Cambridgeshire
Llyn Penrhyn	SH 3077	Gwynedd
Llyn Traffwll	SH 3276	Gwynedd
Loch Bee	NF 7743	Western Isles
Loch Calder	ND 0760	Highland
Loch Eye	TH 8379	Highland
Loch Heilen	ND 2568	Highland
Loch Ken	NX 6870	Dumfries & Galloway

Site	1 km square	County
Loch Leven	NO 1401	Tayside
Loch Linnhe	NM 9862	Highland
Loch Mahaick	NN 7006	Central
Loch Na Keal	NM 5038	Strathclyde
Loch of Boardhouse	HY 2725	Orkney
Loch of Harray	HY 2915	Orkney
Loch of Kinnordy	NO 3655	Tayside
Loch of Lintrathen	NO 2754	Tayside
Loch of Skene	NJ 7807	Grampian
Loch of Strathbeg	NK 0758	Grampian
Loch Quien	NS 0659	Strathclyde
Loch Watten	ND 2256	Highland
Loch Scarmclate	ND 1859	Highland
Loch Spynie	HU 3716	Shetland
Loch Tullybelton	NO 0034	Tayside
Loe Pool	SW 6424	Cornwall
Loughs Neagh & Beg	J 0575	Down/Antrim/Derry/Tyrone/Armagh
Lour	NO 4746	Tayside
Machrie Bay, Arran	NR 8933	Strathclyde
Machrihanish	NS 6922	Strathclyde
Maidens Harbour/Turnberry	NS 1902	Strathclyde
Martin Mere	SD 4105	Lancashire
Mere Sands Wood	SD 4415	Lancashire
Merryton Ponds	NS 7654	Strathclyde
Middle Yare Marshes	TG 3504	Norfolk
Minnis Bay to Reculver	TR 2569	Kent
Minsmere	TM 4666	Suffolk
Morfa Bychan Pools	SH 5537	Gwynedd
Murcar, Aberdeen	NJ 9510	Grampian
Nene Washes	TF 3300	Cambridgeshire
Netherfield Gravel Pit	SK 6339	Nottinghamshire
North Warren	TM 4658	Suffolk
Nosterfield Quarry	SE 2780	North Yorkshire
Ogston Reservoir	SK 3760	Derbyshire
Ormesby Broads	TG 4614	Norfolk
Ouse Washes	TL 5394	Cambridgeshire
Pannel Valley	TQ 8815	East Sussex
Pawston Lake	NT 8632	Northumberland
Pensthorpe Lakes	TF 9428	Norfolk
Pentney Gravel Pits	TF 7013	Norfolk
Pitsford Reservoir	SP 7669	Northamptonshire
Port Meadow	SP 4908	Oxfordshire
Pulborough Levels	TQ 0416	West Sussex
Queen Elizabeth II Reservoir	TQ 1167	Surrey
Queen Mary Reservoir	TQ 0769	Surrey
Queen Mother Reservoir	TQ 0076	Berkshire
Ranworth and Cockshoot Broads	TG 2515	Norfolk
Rhunahaorine	NR 7049	Strathclyde
River Avon: Blashford to Hucklesbrook	SU 1408	Hampshire
River Avon: Fordingbridge	SU 1617	Hampshire
River Avon: Ringwood	SU 1408	Hampshire
River Eden: Rockcliffe to Armathwaite	NY 4758	Cumbria
River Lune: Arkholme to Whittington	SD 5871	Lancashire
River Lune: Caton to Hornby	SD 5566	Lancashire
River Severn: Shrewsbury	SJ 4815	Shropshire
River Soar: Leicester	SK 5805	Leicestershire
River Tay: Perth	NO 1125	Tayside
River Teviot: Nisbet	NT 6725	Borders
River Tweed: Kelso to Coldstream	NT 7737	Borders
River Tyne: Corbridge to Blaydon	NZ 1064	Northumberland/Tyne & Wear
Rookery Pit	TL 0141	Bedfordshire
Rostherne Mere	SJ 7484	Cheshire

Site	1 km square	County
Ruslands Pool	SD 3486	Cumbria
Rutland Water	SK 9207	Leicestershire
Ryton Willows	NZ 1462	Tyne & Wear
St Benets Levels	TG 3815	Norfolk
Saintear Loch	HY 4347	Orkney
Sandbach Flashes	SJ 7259	Cheshire
Scarmclate	ND 1959	Highland
Seahouses to Budle Point	NU 2231	Northumberland
Shibdon Pond	NZ 1962	Tyne & Wear
Slains Lochs/Ythan Estuary	NK 0230	Grampian
Somerset Levels	ST 4040	Somerset
South Forty Foot Drain	TF 2843	Lincolnshire
South Muskham & North Newark Gravel Pits	SK 7956	Nottinghamshire
Stanford Reservoir	SP 6080	Leicestershire
Stantling Craigs and Bunting Craigs Reservoirs	NT 4339	Borders
Stanwick Gravel Pits	SP 9773	Northamptonshire
Stodmarsh	TR 2061	Kent
Stranraer Lochs	NX 1161	Dumfries & Galloway
Stratfield Saye	SU 7061	Hampshire
Strumpshaw Fen	TG 4306	Norfolk
Sutton Bingham Reservoir	ST 5410	Somerset
Swillington Ings	SE 3828	West Yorkshire
Swithland Reservoir	SK 5513	Leicestershire
Tabley Mere	SJ 7276	Cheshire
Talkin Tarn	NY 5458	Cumbria
Tay/Ilsa Valley	NO 1438	Tayside
Tealham and Tadham Moor	ST 4145	Somerset
Tentsmuir	NO 5024	Fife
Thorpe Water Park	TQ 0268	Surrey
Thrapston Gravel Pit	SP 9979	Northamptonshire
Tophill Low Reservoirs	TA 0748	Humberside
Twyford Gravel Pit	SU 7875	Berkshire
Upper Glendevon Reservoir	NN 9004	Tayside
Upper Loch Erne	H 3231	Fermanagh
Upper Quoile	J 4846	Down
Upton Warren	SO 9367	Hereford & Worcester
Walland Marsh	TQ 9824	Kent
Walmore Common	SO 7425	Gloucestershire
Waltham Brooks	TQ 0112	East Sussex
Walthamstow Reservoir	TQ 3589	Greater London
Walton Reservoirs	TQ 1268	Surrey
Washington WWT	NZ 3356	Tyne & Wear
Watch Water Reservoir	NT 6656	Borders
Water Sound	ND 4694	Orkney
Wayoh Reservoir	SD 7301	Lancashire
Westfield Marshes	ND 0664	Highland
Westhay Heath	ST 4142	Somerset
Westhay Moor	ST 4544	Somerset
West Sedgemoor	ST 3525	Somerset
West Water Reservoir	NT 1252	Borders
Whitemore Reservoir	SD 8473	Lancashire
Whittledene Reservoirs	NZ 0667	Northumberland
Willington Gravel Pits	SK 2828	Derbyshire
Windermere	SD 3995	Cumbria
Woolston Eyes	SJ 6588	Cheshire
Wraysbury Gravel Pits	TQ 0073	Berkshire
Yare Valley	TG 3504	Norfolk
Ynys-hir	SN 6896	Dyfed

Appendix 3. TOTAL NUMBERS OF WATERFOWL RECORDED BY WeBS IN ENGLAND DURING WINTER 1992-93.

Wildfowl at all sites	Sep	Oct	Nov	Dec	Jan	Feb	Mar
Number of sites counted	1,243	1,329	1,352	1,365	1,431	1,361	1,371
Red-throated Diver	22	66	73	580	194	98	272
Black-throated Diver	4	0	3	5	3	0	2
Great Northern Diver	0	0	4	4	7	6	6
Little Grebe	2,160	2,138	1,912	1,604	1,411	1,695	1,827
Great Crested Grebe	8,005	8,766	7,852	6,891	4,982	6,945	8,032
Red-necked Grebe	7	10	12	9	11	12	19
Slavonian Grebe	1	10	30	43	36	71	55
Black-necked Grebe	19	11	7	36	35	34	33
Cormorant	8,959	10,202	8,840	9,228	9,392	8,150	8,705
Grey Heron	2,563	2,399	2,041	1,945	1,746	1,993	2,119
Mute Swan	10,496	10,769	10,970	10,446	10,838	9,906	9,554
Bewick's Swan	0	45	2,642	1,052	6,763	6,996	572
Whooper Swan	0	131	1,400	1,447	1,083	1,571	641
Bean Goose	0	0	0	2	352	7	36
Pink-footed Goose	222	+16,448	+37,026	42,246	24,545	19,292	15,182
European White-fronted Goose	13	87	625	1,523	1,818	1,729	3,075
Greenland White-fronted Goose	0	2	1	1	1	3	13
Lesser White-fronted Goose	1	0	1	0	0	2	1
Greylag Goose*	12,718	13,430	11,310	12,928	12,730	9,275	9,052
Snow Goose	11	68	64	35	55	69	35
Canada Goose	35,024	34,905	33,625	38,046	33,615	26,914	21,700
Barnacle Goose	179	311	7,915	274	4,288	11,376	1,512
Dark-bellied Brent Goose	2,656	59,833	93,319	96,563	94,937	94,692	86,631
Light-bellied Brent Goose	380	1,305	+1,762	1,175	1,788	183	8
Red-breasted Goose	1	1	0	0	0	1	0
Egyptian Goose	153	108	76	63	43	51	64
Shelduck	23,503	47,308	50,816	63,634	62,073	56,662	55,734
Mandarin	101	156	136	165	139	116	101
Wigeon	16,791	124,384	144,411	210,857	236,794	126,798	130,283
American Wigeon	0	0	0	0	0	0	2
Gadwall	5,310	7,316	7,598	7,770	7,408	6,828	3,818
Teal	47,360	63,766	76,089	89,110	81,200	64,223	32,302
Mallard	116,498	113,722	123,877	131,255	116,886	79,959	52,662
Pintail	4,524	18,737	13,993	18,075	17,246	11,067	4,982
Garganey	25	2	0	1	3	2	12
Shoveler	6,815	6,792	6,750	7,317	5,813	6,347	5,435
Red-crested Pochard	51	78	134	84	127	74	102
Pochard	11,036	19,225	27,112	28,053	31,416	27,072	10,645
Ferruginous Duck	0	0	2	1	1	1	0
Ring-necked Duck	0	1	2	2	5	0	2
Tufted Duck	27,436	31,568	37,056	41,806	39,939	34,675	29,661
Scaup	63	148	340	202	252	658	289
Eider	8,577	10,510	10,546	4,511	4,489	4,055	10,060
Long-tailed Duck	0	12	98	114	58	98	135
Common Scoter	278	551	536	1,014	462	105	637
Velvet Scoter	2	57	30	60	6	0	4
Goldeneye	20	837	4,654	5,624	6,636	6,701	6,030
Smew	0	0	9	39	78	95	13
Red-breasted Merganser	504	1,198	1,981	1,829	1,565	1,592	1,984
Goosander	320	516	1,094	1,675	1,989	1,586	1,470
Ruddy Duck	1,452	1,899	1,793	2,215	1,917	1,825	1,924
Water Rail	64	142	156	137	107	123	136
Moorhen	5,862	7,000	6,779	5,907	6,950	6,691	7,125
Coot	70,716	79,806	86,342	79,949	75,601	50,298	37,742
TOTAL WILDFOWL**	**436,511**	**703,436**	**823,337**	**931,521**	**908,580**	**687,563**	**568,940**

Waders at estuarine/coastal sites	Nov	Dec	Jan	Feb	Mar
Number of sites counted	86	87	85	87	85
Oystercatcher	244,577	205,017	235,769	141,314	127,194
Avocet	1,769	1,950	1,851	2,137	1,392
Ringed Plover	8,646	7,803	6,302	7,280	5,330
Kentish Plover	0	1	1	1	0
Golden Plover	50,949	98,863	43,533	67,993	45,515
Grey Plover	35,813	37,120	36,490	33,747	37,142
Lapwing	163,134	344,036	115,416	199,473	74,430
Knot	285,892	303,674	282,557	151,441	157,887
Sanderling	3,879	5,951	4,503	3,288	4,314
Little Stint	3	3	1	0	0
Curlew Sandpiper	1	0	0	0	0
Purple Sandpiper	783	813	1,022	1,200	1,281
Dunlin	334,359	418,034	374,193	328,440	295,726
Ruff	115	134	45	148	161
Jack Snipe	31	18	18	23	11
Snipe	2,349	1,823	1,049	1,204	1,198
Woodcock	5	0	2	0	0
Black-tailed Godwit	9,740	8,720	6,178	5,862	7,954
Bar-tailed Godwit	37,412	30,972	31,443	24,094	32,273
Whimbrel	0	4	5	2	8
Curlew	48,274	62,235	44,351	46,363	47,952
Spotted Redshank	63	57	44	48	59
Redshank	55,273	56,120	50,628	44,285	47,233
Greenshank	118	161	99	120	92
Green Sandpiper	36	29	22	41	33
Common Sandpiper	18	11	19	17	12
Turnstone	12,665	11,301	11,883	10,207	10,906
TOTAL WADERS	1,295,904	1,594,850	1,247,424	1,068,728	898,103
TOTAL WATERFOWL*	2,121,282	2,528,316	2,157,750	1,758,284	1,469,162
Kingfisher (all sites) 263 255	200	163	122	148	190

+ Counts include data from the following goose censuses: national census of Pink-footed in October and November; December census of Dark-bellied Brent Geese; November census of Light-bellied Brent Geese on Lindisfarne; See *Progress and Developments* and *Species Accounts* for more details.

* Comprises mainly feral birds, and small numbers of the Icelandic breeding population.
** Total wildfowl represents numbers of all divers, grebes, Cormorant, swans, geese, ducks and rails.
*** Total waterfowl represents numbers of all wildfowl (as above), waders at estuarine/coastal sites and Grey Heron.

Footnote: Presentation of wildfowl and wader totals differ slightly in this Appendix. Where a WeBS site crosses a country boundary (e.g. The Severn Estuary), only wildfowl within the English part of the site are included in the above table. However, for waders, the total counts of birds on the site, in both England and the adjacent country, are included in the above table.

Appendix 4. TOTAL NUMBERS OF WATERFOWL RECORDED BY WeBS IN SCOTLAND DURING WINTER 1992-93

Wildfowl at all sites	Sep	Oct	Nov	Dec	Jan	Feb	Mar
Number of sites counted	379	450	403	431	426	459	396
Red-throated Diver	851	201	256	266	204	151	
Black-throated Diver	0	3	20	14	3	14	27
Great Northern Diver	5	9	16	20	17	33	20
Little Grebe	343	244	146	139	99	142	140
Great Crested Grebe	1,089	716	397	449	375	1,183	502
Red-necked Grebe	14	5	6	2	9	22	4
Slavonian Grebe	16	47	49	60	50	92	88
Black-necked Grebe	1	0	0	2	2	2	6
Cormorant	2,446	3,054	2,341	2,490	2,166	2,915	1,611
Grey Heron	415	539	369	394	237	303	218
Mute Swan	2,368	2,790	2,408	2,616	2,352	2,463	1,897
Bewick's Swan	0	9	15	14	0	5	0
Whooper Swan	7	1,116	1,739	1,171	916	933	1,028
Bean Goose	0	2	0	0	0	1	3
Pink-footed Goose	12,641	+181,413	+130,486	67,735	54,630	75,736	50,084
European White-fronted Goose	15	19	15	6	1	12	13
Greenland White-fronted Goose	0	871	+13,846	268	70	400	+15,097
Greylag Goose*	1,007	+88,533	+91,680	45,356	30,183	22,070	16,578
Snow Goose	2	2	3	2	6	7	5
Canada Goose	618	125	146	245	177	127	82
Barnacle Goose	11	5,081	2,679	+39,150	1,381	4,195	+27,733
Dark-bellied Brent Goose	0	0	0	0	1	0	4
Light-bellied Brent Goose	2	6	5	4	8	4	7
Shelduck	5,393	2,040	3,092	4,518	5,459	7,350	3,914
Mandarin	0	3	5	4	2	3	2
Wigeon	7,125	55,976	32,979	62,172	48,158	38,173	18,406
American Wigeon	0	1	3	1	2	1	0
Gadwall	309	157	76	71	25	65	55
Teal	8,258	12,490	10,918	13,465	10,678	9,594	3,812
Mallard	18,872	28,716	27,302	32,672	26,531	21,005	10,607
Pintail	647	1,335	772	739	922	1,024	242
Shoveler	661	887	561	209	84	113	168
Red-crested Pochard	2	0	0	1	0	0	0
Pochard	3,077	5,093	5,161	4,740	4,178	2,982	1,097
Ring-necked Duck	0	0	2	2	0	0	0
Tufted Duck	9,312	8,516	8,503	9,422	9,393	7,890	5,659
Scaup	217	1,167	1,647	2,078	3,083	3,070	2,095
Eider	18,990	10,905	11,222	10,053	12,315	16,957	15,249
King Eider	0	1	0	0	0	0	0
Long-tailed Duck	5	404	584	854	1,140	1,098	893
Common Scoter	1,180	1,467	673	669	1,744	2,087	869
Velvet Scoter	28	58	175	202	184	265	290
Surf Scoter	0	0	3	1	6	2	2
Goldeneye	222	2,177	4,192	6,336	7,862	9,159	6,763
Smew	0	0	4	9	6	6	5
Red-breasted Merganser	1,554	1,872	1,153	1,184	1,078	2,222	1,147
Goosander	543	934	732	654	841	765	400
Ruddy Duck	21	22	4	1	10	1	12
Water Rail	9	8	8	5	0	3	10
Moorhen	551	736	631	496	386	516	554
Coot	5,992	7,143	7,345	6,468	5,824	4,030	3,226
TOTAL WILDFOWL**	104,746	427,748	365,110	317,899	233,211	240,143	191,647

Waders at estuarine/coastal sites	Nov	Dec	Jan	Feb	Mar
Number of sites counted	40	44	40	43	32
Oystercatcher	56,209	59,335	57,197	66,248	37,739
Ringed Plover	1,562	1,621	1,795	1,825	851
Golden Plover	10,072	16,688	4,686	7,271	4,934
Grey Plover	1,846	1,700	2,647	1,827	1,854
Lapwing	21,177	28,554	17,236	18,824	3,632
Knot	6,437	9,594	27,889	17,831	9,053
Sanderling	312	224	419	482	194
Purple Sandpiper	659	638	585	437	438
Dunlin	16,697	29,644	37,516	31,284	17,495
Ruff	1	1	7	1	0
Jack Snipe	5	2	1	1	1
Snipe	210	74	73	99	73
Black-tailed Godwit	152	176	176	175	145
Bar-tailed Godwit	4,213	7,850	10,219	11,323	3,930
Curlew	10,220	16,930	13,131	21,905	10,038
Spotted Redshank	1	0	0	0	0
Redshank	14,875	15,958	15,472	17,791	13,649
Greenshank	17	26	14	22	17
Green Sandpiper	0	1	0	0	0
Turnstone	2,491	2,688	3,764	3,022	2,614
Grey Phalarope	0	0	1	0	0
TOTAL WADERS	**147,156**	**191,704**	**192,828**	**200,368**	**106,657**
TOTAL WATERFOWL*	**512,635**	**509,997**	**426,276**	**440,814**	**298,522**
Kingfisher (at all sites) 8 4	6	6	4	3	5

+ Counts include data from the following goose censuses: national census of Pink-footed and Greylag Geese in October and November; December and March censuses of Barnacle Geese on Islay; December census of Barnacle Geese on the Solway; international censuses of Greenland White-fronted Geese in November/December and March/April. See *Progress and Developments* and *Species Accounts* for more details.

* Comprises mainly birds from the Icelandic breeding population, with up to 2,340 feral birds (Delany 1992).
** Total wildfowl represents numbers of all divers, grebes, Cormorant, swans, geese, ducks and rails.
*** Total waterfowl represents numbers of all wildfowl (as above), waders at estuarine/coastal sites and Grey Heron.

Footnote: Presentation of wildfowl and wader totals differs slightly in this Appendix. Where a WeBS site crosses a country boundary (e.g. The Solway Estuary), only wildfowl within the Scottish part of the site are included in the above table. However, for waders, the total counts of birds on the site, in both Scotland and the adjacent country, are included in the above table.

Appendix 5. TOTAL NUMBERS OF WATERFOWL RECORDED BY WeBS IN WALES DURING WINTER 1992-93

Wildfowl at all sites	Sep	Oct	Nov	Dec	Jan	Feb	Mar
Number of sites counted	125	133	146	140	159	163	158
Red-throated Diver	0	15	6	4	1	4	72
Great Northern Diver	0	0	1	0	2	1	2
Little Grebe	97	151	172	134	121	130	105
Great Crested Grebe	138	98	158	107	68	141	186
Red-necked Grebe	0	2	1	1	1	0	1
Slavonian Grebe	0	0	0	0	3	2	0
Black-necked Grebe	0	0	1	0	0	0	0
Cormorant	770	675	481	479	445	530	460
Grey Heron	142	159	106	103	76	172	141
Mute Swan	385	285	310	251	277	192	287
Bewick's Swan	0	0	11	5	11	15	0
Whooper Swan	0	5	85	66	62	53	103
Pink-footed Goose	0	0	2	1	2	2	1
Greenland White-fronted Goose	0	67	122	122	134	124	124
Greylag Goose: feral	232	169	139	365	433	357	350
Canada Goose	643	407	763	802	1,193	744	435
Barnacle Goose	0	0	5	1	1	9	1
Dark-bellied Brent Goose	36	351	536	500	364	554	974
Light-bellied Bent Goose	84	9	23	17	27	26	10
Egyptian Goose	0	0	1	0	1	0	1
Shelduck	195	1,462	3,521	4,562	4,339	4,899	4,477
Mandarin	1	0	0	1	1	0	4
Wigeon	1,190	7,673	10,884	12,677	14,304	5,686	3,649
American Wigeon	0	0	0	0	0	0	1
Gadwall	44	61	88	62	55	75	70
Teal	819	4,093	5,911	7,352	7,232	4,479	1,612
Mallard	6,560	6,975	7,126	6,577	5,882	4,260	2,686
Pintail	77	377	767	2,179	1,404	1,117	256
Shoveler	60	106	219	347	285	365	178
Red-crested Pochard	1	0	0	1	2	2	2
Pochard	318	836	1,245	1,103	881	1,255	452
Tufted Duck	765	1,005	1,195	1,401	757	1,262	629
Scaup	8	4	8	11	125	6	0
Eider	0	1	22	0	5	3	5
Long-tailed Duck	0	0	0	1	0	1	0
Common Scoter	81	160	101	172	69	101	157
Goldeneye	831	310	269	343	528	381	
Smew	0	0	1	1	8	7	15
Red-breasted Merganser	133	273	192	124	135	219	286
Goosander	43	64	144	81	81	129	301
Ruddy Duck	118	144	115	37	79	171	105
Water Rail	4	11	6	22	7	7	18
Moorhen	161	146	156	159	126	145	177
Coot	2,736	2,192	1,615	1,981	2,221	1,644	1,117
TOTAL WILDFOWL*	**16,023**	**28,196**	**36,729**	**42,275**	**41,710**	**29,586**	**20,123**

Waders at estuarine/coastal sites	Nov	Dec	Jan	Feb	Mar
Number of sites counted	22	23	22	25	24
Oystercatcher	97,736	61,244	91,970	29,835	19,493
Avocet	7	0	0	0	0
Ringed Plover	860	869	612	856	369
Golden Plover	8,449	3,327	9,373	7,145	3,463
Grey Plover	1,766	1,051	2,532	1,339	950
Lapwing	17,066	13,201	22,207	11,365	4,131
Knot	15,077	4,358	37,954	9,356	8,367
Sanderling	592	214	609	506	496
Little Stint	3	1	0	0	0
Purple Sandpiper	27	54	42	4	49
Dunlin	50,937	65,041	42,858	57,756	18,857
Ruff	1	0	4	1	6
Jack Snipe	17	8	9	10	8
Snipe	484	338	206	131	91
Woodcock	0	0	0	0	1
Black-tailed Godwit	1,837	1,552	1,744	109	632
Bar-tailed Godwit	361	410	1,379	1,541	187
Whimbrel	0	0	15	0	0
Curlew	11,076	15,779	13,916	11,710	9,894
Spotted Redshank	13	7	10	5	12
Redshank	12,247	11,905	8,002	4,337	7,389
Greenshank	38	43	26	38	14
Green Sandpiper	2	4	2	5	1
Common Sandpiper	3	3	2	2	1
Turnstone	1,659	1,381	871	1,438	1,360
TOTAL WADERS	220,258	180,790	234,343	137,489	75,771
TOTAL WATERFOWL**	257,093	223,168	276,129	167,247	96,035
Kingfisher (at all sites) 8 15	8	10	4	8	18

* Total wildfowl represents numbers of all divers, grebes, Cormorant, swans, geese, ducks and rails.
** Total waterfowl represents numbers of all wildfowl (as above), waders at estuarine/coastal sites and Grey Heron.

Footnote: Presentation of wildfowl and wader totals differs slightly in this Appendix. Where a WeBS site crosses a country boundary (e.g. the Severn Estuary), only wildfowl within the Welsh part of the site are included in the above table. However, for waders, the total counts of birds on the site, in both Wales and the adjacent country, are included in the above table.

Appendix 6. TOTAL NUMBERS OF WATERFOWL RECORDED BY WeBS IN THE ISLE OF MAN DURING WINTER 1992-93.

Wildfowl at all sites	Sep	Oct	Nov	Dec	Jan	Feb	Mar
Number of sites counted	9	10	10	10	10	10	9
Little Grebe	0	1	0	0	0	1	0
Cormorant	0	0	0	1	0	0	0
Grey Heron	3	1	1	1	1	1	0
Whooper Swan	0	0	1	14	23	24	21
Pink-footed Goose	0	1	0	0	0	0	0
Greylag Goose	0	0	1	4	0	0	1
Canada Goose	0	0	11	11	4	3	6
Shelduck	0	0	0	0	7	4	5
Wigeon	14	86	127	162	410	295	4
Teal	15	52	77	121	141	30	37
Mallard	330	423	557	450	453	230	113
Pintail	0	0	0	0	0	0	1
Shoveler	0	0	0	0	4	12	2
Pochard	2	1	4	5	15	9	3
Tufted Duck	13	14	10	10	16	8	3
Scaup	1	0	0	0	0	0	0
Goldeneye	0	4	7	1	3	0	0
Moorhen	1	2	2	18	6	9	10
Coot	16	17	17	18	22	33	36
TOTAL WILDFOWL*	**396**	**604**	**817**	**834**	**1,111**	**668**	**252**

* Total wildfowl represents numbers of all divers, grebes, Cormorant, swans, geese, ducks and rails.

Footnote: No counts of waders at estuarine/coastal sites (under the BoEE) were made on the Isle of Man in 1992-93.

Appendix 7. TOTAL NUMBERS OF WATERFOWL RECORDED BY WeBS IN THE CHANNEL ISLANDS DURING WINTER 1992-93.

Wildfowl at all sites	Sep	Oct	Nov	Dec	Jan	Feb	Mar
Number of sites counted	1	10	10	9	10	9	10
Little Grebe	0	2	0	3	3	2	2
Great Crested Grebe	0	0	0	0	1	0	3
Slavonian Grebe	0	0	0	0	0	2	0
Cormorant	22	15	28	11	15	6	17
Grey Heron	17	82	6	16	35	16	7
Mute Swan	0	1	1	1	0	0	0
Pink-footed Goose	0	2	0	0	2	0	2
Dark-bellied Brent Goose	1	4	18	10	2	52	85
Mandarin	0	0	1	2	0	0	0
Wigeon	0	0	1	1	2	0	0
Gadwall	0	0	0	0	2	5	2
Teal	0	23	22	20	127	79	25
Mallard	0	201	229	274	228	192	151
Pintail	0	0	0	0	2	1	0
Shoveler	0	2	9	0	31	23	32
Pochard	0	0	10	0	10	8	7
Tufted Duck	0	94	94	83	106	27	68
Red-breasted Merganser	0	0	0	0	0	1	0
Water Rail	0	13	26	13	14	10	2
Moorhen	0	102	59	79	103	81	85
Coot	0	66	69	40	71	27	20
TOTAL WILDFOWL*	**40**	**722**	**658**	**645**	**873**	**625**	**595**

Waders at estuarine/coastal sites	Sep	Oct	Nov	Dec	Jan	Feb	Mar
Number of sites counted			1	1	2	2	1
Oystercatcher			431	2,370	1,864	1,347	400
Ringed Plover			247	558	315	576	48
Golden Plover			0	2	0	0	0
Grey Plover			83	497	425	777	103
Lapwing			0	1	100	0	0
Sanderling			38	257	282	248	6
Dunlin			525	3,862	2,962	2,227	132
Common Snipe			0	0	7	0	0
Bar-tailed Godwit			0	92	193	191	2
Curlew			88	374	484	308	72
Redshank			71	255	409	192	27
Greenshank			0	16	8	5	0
Common Sandpiper			0	1	0	0	0
Turnstone			562	460	702	786	565
TOTAL WADERS			**2,045**	**8,745**	**7,751**	**6,657**	**1,355**
TOTAL WATERFOWL**			**2,709**	**9,390**	**8,624**	**7,282**	**1,950**
Kingfisher (at all sites)	0	0	0	0	2	0	0

* Total wildfowl represents numbers of all divers, grebes, Cormorant, swans, geese, ducks and rails.
** Total waterfowl represents numbers of all wildfowl (as above), waders at estuarine/coastal sites and Grey Heron.